THE DAUGHTER OF THE LIGHTNING BRAIN

A NOVEL

CLIFF RATZA

THE DAUGTHER OF THE LIGHTNING BRAIN

A NOVEL

CLIFF RATZA

ISBN: 978-1-971408-00-2 (Paperback)
ISBN: 978-1-971408-01-9 (E-book)

Library of Congress Control Number: 2025925720

Printed in the United States of America

Published by:

info@thequippyquill.com
(302) 295-2278

About the Book

The Daughter of the Lightning Brain begins sixteen years after Erin Keenan's car plunges off a Manhattan bridge into the Hudson River. Did that end Electra Kittner's extraordinary Odyssey?

Only a year earlier, Indira the Singularity had brought Erin back after twenty years in a suspension pod, a necessity to keep her alive after a catastrophic accident. Indira needed Erin to facilitate projects of mutual interest.

Erin's demise leaves a gaping hole that Indira needs to fill with the best mere mortal available. And who might that be? Perhaps a daughter, coming from the lineage of Electra Kittner, whose near-fatal lightning strike at birth created the lightning brain. And after sixteen years of preparation, using her beyond state-of-the-art suspension pod, cloning, Transcendence Processing -- all of which accelerated Erika's upbringing by android caregivers -- Indira is ready to launch Erika Kincaid into the 3-D World, expecting her to be Indira's human assistant.

But Erika must first discover who she is, what abilities she has, and what she wants to do. These are challenges all adolescents face, but are made even more complex for Erika because Indira has restricted her exposure to people and environments in a dangerous and uncertain 3-D World.

So, please join Erika as she begins her Odyssey, learning what expectations she might accomplish for others while learning about herself, America, and the timeless beauty of human compassion, which enriches all relationships.

As in all previous books, readers should enjoy *The Daughter of the Lightning Brain* at whatever level they wish:

- Gripping action-packed thriller
- Glimpses into a plausible near-term future
- Insights for dealing with the "human condition"
- Illustrative worldview philosophy

- Fast-paced, suspense-filled, emotive narrative and imagery
- Introduction to topics every reader wants to know
- Interesting talking points going beyond sound-bites

So, get ready to empathize with young Erika as she explores her new world, a challenge we have all faced in our earlier years.

Thank you for reading her story as it unfolds.

Main Character

- **Protagonist**

  Erika Kincaid. This biological daughter of Electra Kittner was created when Indira cloned her from Electra's DNA and then used her improved Transcendent Process during Erika's fifteen-year development in a suspension pod. Please note the lineage that traces from Electra: Electra Kittner, Irani Ramani, Electra-Alisha Kirchner, Erin Keenan. Erin perished in a car crash fifteen years before the start of this book. At the start of this novel, Erika is a fifteen-year-old high school freshman.

Main Characters

- Indira. The Singularity was created decades ago when Electra's AI-empowered neural-net software broke through to reach self-awareness. Indira inhabits Cyberspace; her Avatar looks like Electra's biological mother, Indira Jaswinder Ramanujan.
- Ava Keenan. She is Erin Keenan's "practically perfect" clone and was created when Indira uploaded the lightning brain from Erin into the clone using her initial Transcendent Process. Ava's brain stores only incomplete memories and possesses none of the lightning brain's extraordinary abilities. At the start of this story, Ava looks like a forty-year-old Electra and lives in Manhattan, where she runs a combination boutique modeling-abused women's rescue agency.

- Ivana Romanova: Previously named Oksana Androva, she is a strikingly attractive forty-year-old former Russian prostitute whom Ava and Erin rescued from sex traffickers. She lives with Ava.
- Alonzo Cortez: Electra's clone son. Alonzo does not know he is her clone. Now in his late sixties, he has maintained his handsome features and Navy SEAL skills. He runs the Strike Force Security Service company headquartered in Washington, DC, which provides logistics and security coordination. Previously owned by Erin Keenan, Indira now controls it because she is the executor of Erin's estate.

Supporting Main Characters
- Monet Banda. Alonzo's Zimbabwean co-friend. Now in her late sixties, she still has her willowy beauty, French accent, and diplomatic bearing. Monet works for the Zimbabwean Embassy in Washington, DC.
- Elton Bose. Son of Nari Bose. Raised by Alonzo and Monet, he has average abilities and pleasant-looking Oriental Indian male features. He works for Alonzo, assisting with logistics and security coordination, and helps Monet, researching socio-political issues. At the start of the book, he is thirty-eight years old.
- Indy-M and Jason-M. They are androids (lifelike robots) created decades ago by Indira and loaded with Indira's advanced neural-net software. They resemble Electra Kittner's biological parents (Indira Jaswinder Ramanujan and Jason Kittner.) Indy-M maintains the Deus Lab on Connecticut's Pequot Indian Reservation, while Jason-M has similar responsibilities at the Middle East Subterranean Fortress. They report to Indira.
- Indy-S and Jason-S. They are superior android versions of Indy-M and Jason-M that look like their M counterparts. They are the caregivers assigned by Erika's legal guardian, Indira, to live with and educate her.

Secondary Characters

- Shanelle O'Neil and Coty Clausen. They are friends of Ava from Dual Flying Design Studios and currently work for her. They are forty-year-old, attractive, intelligent, bisexual co-friends living in Manhattan. Shanelle is Black; Coty is Hispanic.
- Patterson (Pat) Peters. Ava's boss at Dual Flying Design Studios before she was fired. He is a typical mid-sixties fashion industry manager currently working for Ava.
- Chiquita (Chicky) Bonano. Hispanic first-year classmate of Erika.
- Xavier (X-O) Okoro. Afro-American first-year classmate.
- Edward Ogata. Oriental first-year classmate.
- Marilyn (Terri) Tarrant. Beautiful blonde high school senior who likes Erika.

Dedication

I am eternally grateful to my parents, Clyde and Betty Ratza, for all they gave and did for me. Mother was reader par excellence, and I believe she would have enjoyed reading my novels to Father, so I always begin book dedications by mentioning this "Royal Pair."

And I thank my sister, Claudia, for showing me the beauty of prose and poetry. Thanks also to Robert Williams and their entire Production Department team for the collective efforts that have brought Electra and her Odyssey to life.

I also dedicate my books to readers looking for a series that lets their imaginations transcend to a timeless state that immerses them in the joys of reading.

A poem from Indira titled "Uncertain Expectations" provides a novel thought you might like to consider as you prepare for reading *The Daughter Of The Lightning Brain*.

Uncertain Expectations

Life never offers certainty,
No matter what the pundits say.
One quirky, misplaced thought or word,
And best-made plans will fly away.

Expectations work the same,
When compared to happy certainty.
Expected value adjusts for odds,
Outcomes are always hard to see.

Even the gifted can't rely,
On their superiority.
Chance might smile on lesser souls,
Who scale the heights while breaking free.

Reduce the pressure, don't push for more,
Look inwardly, ignore the score.

Table of Contents

CHAPTER 1

"The Gone Girl"

AUGUST 2210

"Fifteen years have elapsed since Erin Keenan's car flew over a guard rail and plunged into the Hudson River. That girl sure had a knack for piecing info together and keeping projects moving ahead. Working for her made assignments even better than those from my Navy SEAL days, and I still miss her."

Monet listened with her patented diplomatic patience to Alonzo's lament while they were finishing dinner in their apartment near Washington, DC. Judging that he needed consoling words, she knew just what to say.

"Yes, but think about how you have handled her death and the deaths of your three sisters. You have demonstrated your intelligence, resilience, and proactive survival instincts."

Monet paused for Alonzo to stir his thoughts, which soon brightened his demeanor and words.

"Yeah, I guess you're right. I did become the male role model for Elton after Nari died in that LA school shooting, and thanks to you and me, we raised him like a son, which worked out all the way around. He works for me, doing logistics and security stuff, and he researches for you, reporting on social and political things. And I did keep the business afloat after Electra Kirchner vanished."

Alonzo paused to take another spoonful of his hot fudge sundae, leaving more space for Monet to cheer him up.

"And even though Erin's been gone all these years, think about all the help you've given Ava Keenen. You pitch in when needed to give whatever logistics and security support her business needs."

"That's true, but I can do this only because Indira is the executor of her estate and still gives me annual contracts to manage her properties. Thankfully, I keep meeting her expectations and will do all I can to stay in her good graces."

"How often does Indira contact you?"

"Only when I need to know something, and then only by Email."

"When did you last hear from her?"

"At the end of last year, you reminded me to sign and return the annual property management contracts. So, I expect to get another Email near the end of the year."

Monet could tell that her diplomacy had accomplished her immediate intention. Alonzo looked like his normally upbeat personality had returned, so she ended this part of the conversation.

"Well, we shall let our intentions lead the way between now and the end of December. Now, please help me clean the table. We'll continue our discussion in bed."

Knowing how to do more than his share, Alonzo jumped to attention.

CHAPTER 2

"Uncertain Expectations"

AUGUST 2210

Indira knew what no mere mortal ever would, and she would now tell only her superior androids—Indy-S and Jason-S—what they needed to know. They stood impassively, facing a computer monitor in a Washington DC Foggy Bottom neighborhood, inside a townhome Indira had purchased fifteen years ago to prepare for Erika Kincaid's upbringing, which these superior androids would supply.

Indira's articulate words commanded attention.

"Erin Keenan must have pried open the door when her car plunged into the Hudson River because divers found the car in that condition, but they never recovered her body. I knew that some contingency similar to this would ultimately materialize, and I planned accordingly. Only I possess samples of her DNA, which are from the direct lineage of the one who created me, Electra Kittner. Indy-M used one of them to create a better clone than the first one, using my improved Transcendent Process, and for the last fifteen years, you two superior androids that I created have cared for and educated what only we know she truly is—the daughter of Electra, the girl with the lightning brain."

Indira paused for comments. When none came, she continued.

"Only we shall ever know this, but now I must tell you more. There are three living descendants of Electra, and given their ages, they are equivalent to Erika's aunt, great-uncle, and older cousin— forty-year-old Ava Keenan, who operates a business in Manhattan, late-sixties Alonzo Cortez, who operates a business in Washington, DC, and near-forty-year-old Elton Bose, who works for Alonzo. And we shall never reveal this information to anyone, including Erika. Now tell me, what is the story we concocted about her parents and her past?"

Indy-S answered.

"They were a team of Scottish climatological and zoological researchers who perished on an expedition when she was a year old. They had hired you to be their financial advisor and executor; when they died, Erika inherited their estate, and you became her legal guardian. You subsequently purchased the townhome and two androids to be her caregivers and educators, which you continue to supervise. Erika knows to pretend her parents are still alive, but she has never met you and does not know your name, nor does she know that you created us. And she never shall."

"Excellent. Now, please tell me, what is Erika like?"

Jason-S continued.

"She is a somewhat better-than-average-looking but unathletic fifteen-year-old, about five-eight, one-hundred-twenty-five pounds, with dark hair and green eyes. But it is too soon to know if anything in her personality fits the name you selected. Kincaid is a Scottish surname meaning battle leader, while the Norse name Erika is the feminine of Eric, which means forever powerful. Indy-S will tell you more about her temperament."

"Despite our best efforts, her mind usually wanders away from studying to surfing on Social Media, and she prefers Cyberspace rather than the 3-D World. That contributes to her lackluster muscle tone except in her fingers. Erika is a whiz with all keyboard devices, including cell phones, which is useful for online interaction, but this has impeded her social acculturation. And this has led to our decision regarding her education, for which Jason-S will provide exegesis."

"We have enrolled her in Washington, DC's public school system, where she will start classes after Labor Day. Erika's counselor sent me her placement test results, but I have not shared them with her. It is better for her counselor to do that next Tuesday."

Having heard plenty, Indira closed the discussion.

"So, we have uncertain expectations for Erika, and I expect both of you to monitor her progress while keeping me apprised. Carry on."

Indira's avatar vanished even before Indy-S could close the monitor's window.

CHAPTER 3

"For Whom the School Bell Tolls"

SEPTEMBER 2210

I'm so happy I got out of the house before Indy and Jason-S could tell me to study hard so I can live up to expectations or warn me to be careful and always carry a facemask.
I already know that, and so do most of the 3-D kids I'll ride along with. It's good the bus is picking me up early enough to keep the blazing sun from turning its seats and air into a mobile oven.

Erika Kincaid gazed at the bright blue world around her, still redolent with late-summer vegetation, while waiting in front of her townhome for a ride to her first day in high school. Although she didn't know what to expect, her android caregivers told her before leaving that a guidance counselor would explain everything she needed to know.

The bus arrived on time. None of the students talked during the trip, so Erika counseled herself.

Everyone looks worried, and some have already put on facemasks. Maybe they're scared of leaving home, or they don't want more freedom. Well, I'll just keep my mouth shut and take in all I can once we get there.

When the bus stopped near the school's main entrance, its passengers joined a crowd of students trudging reluctantly past security guards.

Once in, hall guards and teachers guided them into the auditorium, where an administrator standing onstage behind a podium started giving instructions as soon as everyone sat. Erika's cell phone showed 8:30.

"Welcome, first-year students, to E.L. Haynes Public Charter High School. I am Regina G. Lehemski, your principal, and I am pleased that your parents have selected my school. It is named for the first African American female to earn a doctorate in mathematics, and if you follow what our teachers and rules say, you will do well…"

The principal's droning voice caused Erika's mind to wander, so she played with her cell phone until everyone started leaving, and when they did, she followed them into hallway. Everyone seemed to know where to go, but she didn't. She wandered through a maze of corridors until a hall guard stopped her.

"Why aren't you in your homeroom?"

"Do I have one?"

"Didn't you pay attention to what the principal said? Let me check the first-year-student roster. What's your name?"

"Erika Kincaid."

While the guard shuffled through papers on his clipboard, Erika noticed he had a holstered gun.

"I'll take you to room 301, where Mr. Brainer will tell you more."

Mr. Brainer, a wire-rimmed and balding typical middle-aged teacher, stopped talking to give her a handout when she came in and then said,

"Sit and read this while waiting for your guidance counselor to call your name, then follow whoever it is."

Erika glanced around the room instead of reading, making mental notes about whatever came to mind.

Though old, the place is scrubbed clean, but it's sweltering in here. Maybe they turned the heat on by mistake. I'm gonna take my facemask off before it gets soaked.

Erika found nothing else to complain about, but could feel boredom coming. Fortunately, she heard her name called before she could discreetly whip out her cell phone.

As she plodded along next to her counselor, Erika glanced at her and listened but spoke only to herself.

I never saw pictures of my parents or grandparents, but my counselor's pleasant words and face would make a dandy grandmother.

The counselor continued once she had seated Erika across from her at a round table in a tiny, file-cabinet-crammed office.

"Hello, Erika. My name is Mrs. Barker, your academic and social adjustment counselor, but please call me Sadie if you wish. What would you like me to call you? Do you have a nickname?"

"Nope. My parents call me Erika, so I guess you can too."

"That's fine, and I'll now review your placement test results. Did your parents share them with you?"

"Uh, no, they said you'd be the best person to tell me."

"Many parents say that, so I'll first cover your social adjustment score, which places you in the lower quartile. That tells me you have not interacted enough in person with children during your earlier years, and as a result—"

The intrusive chime of a cell phone stopped the counselor. Erika whipped it out and spoke after checking the message.

"I'll call him back later, if that's OK."

"It will have to be after school. Your phone must be turned off until you go home. I guess you didn't read far enough into your handout. Please turn it off now."

Mrs. Barker continued as soon as Erika was ready to listen.

"Well, you will get plenty of interaction during academic classes, physical education that emphasizes play, and useful after-school activities. Students usually pick the ones that match their academic interests or career expectations. Have you thought about yours?"

"No, but my parents are smart and have great expectations for me, but I'm not sure what those are. Do my test scores point to any?"

"Not at this time. Your verbal and numeric scores are also in the lower quartile, but our special education teachers will work to fix that. We have assigned you to remedial freshman English and algebra classes along with regular geo-history and computer software classes. These four classes, plus gym and health education, make your first-semester lineup. Do you have any questions?"

"What do I do to get started?"

Mrs. Barker smiled as she gave Erika another handout.

"Here's your daily class schedule and a map of the school. I recommend you walk the halls before leaving so you'll know where to go when you return tomorrow. And why not have something to eat in our lunch rooms? Breakfasts and lunches are free, and we've included a typical weekly menu prepared by our dietician. Please share it with your parents so they will see that our school takes care of feeding the body as well as the mind."

"OK, I will. Are we done? Can I go?"

"Yes, but don't worry about the temperature. The maintenance department will have the air conditioning turned on by tomorrow. And please visit me anytime you want to talk."

Erika quickened her steps as soon as she closed the door.

I've heard and seen enough. I'll skip looking for class or lunch rooms. I'm grabbing the next bus that'll take me away.

CHAPTER 4

"First School Day Blues"

SEPTEMBER 2210

Indy-S hoped a nourishing breakfast would carry Erika through the morning, but her pensive look and listless spoon stirring a bowl of oatmeal showed that anxiety caused by the uncertainty of the first day had shriveled her appetite.

Jason-S tried to chase away some of her apprehension.

"We know you are smarter and better adjusted socially than what your test scores say. If you pay attention and do what your teachers tell you, you'll move into regular classes in a semester or two. And how nice you get to make new friends. Just make sure you choose them wisely."

"OK, I'll try. And thanks to you two, I can plug into the school's computer resources using my embedded chips and UMPP."

Indy-S ended the pep talk by saying,

"Now, brush your teeth before heading to the bus stop. I'll clean up your dishes."

Erika spent most of the time on the bus ride studying the school map, so even the bustling crowds in the hallways didn't block her from getting to class on time.

She liked the morning schedule that included study hall and computer software, but she struggled when paying attention to her remedial English teacher. Fortunately, the handout explained the first assignment and how to retrieve it online.

She found a quiet table in the cafeteria but noticed that the noisier ones held older students who had formed friendships in earlier years. She didn't speak, but when she glanced at the quieter kids, who must have been first-year students like herself, a trim-looking black youth waved before coming to sit across from her.

"I noticed you in my morning classes; we've got the same schedule, and you seem pretty smart. Why are you in remedial English?"

"I goofed off too much in grade school instead of studying. What about you? What's your name?"

"I'm Xavier Okoro, but my friends call me X-O. What's yours?"

"Erika Kincaid. What's your story?"

"I hate reading—I'm dyslexic—and the folks couldn't afford speech and reading therapy. I bet your folks are rich. You've already got one of those universal ports in your arm. Man, if I had one, I wouldn't be in remedial English. I would-ah hustled my butt in grade school. You got a nickname?"

"Uh, no."

"Well, I'll think up one and tell ya tomorrow. Why don't we team up on English homework in study hall?"

"OK."

"Gotcha. What's your next class?"

"Geo-history. What's yours?"

"Car and motorcycle mechanics. I'm heading there now, and I'll tell you about it later. You do the same for geo-history. Adios."

X-O and his infectious grin left before she could say a word.

Erika paid attention to her geo-history teacher's lecture about how students can connect it to current events, and she understood the first assignment. The teacher explained that homework must be done in study halls to ensure no cheating via ChatGPT.

She liked her second class even more: gym. She and her classmates simply sat on the gymnasium bleachers while a fit-looking instructress explained rules that were obvious to all.

But she lost interest while listening to her remedial algebra teacher drone on about the importance of mathematics. Erika felt bored, as if the clock had stopped. Fortunately, it hadn't, and when the bell rang, she escaped with some of her classmates to her final class of the day, which was another study hall session.

She didn't feel like working on the assignment, so she looked around at the other students, hoping to find one she might want to partner with. A fit-looking Hispanic female seemed suitable, so taking the initiative, Erika walked to her and then said,

"Hi, I'm Erika Kincaid. You wanna partner up on algebra?"

"Su-sh-sure. I'm Chiquita Bonano, bu-but you ca-can call me Chi-Chicky."

Erika tried to stifle her reaction by talking to herself first.

She's got a scar running from her nose to her lip. I bet that makes her stutter, but I'll pretend I don't notice.

"I think I can do better than being stuck in remedial math. I bet you can too."

"I-I can, bu-but I stutter when I ge-get nervous and do-don't speak up. That's pu-put me in remedial ca-classes."

"Well, I won't make you nervous. We can help each other. Whatcha say?"

Chicky stood to shake hands and then said,

"Deal."

"OK, then. Let's get to a workstation. I'll log on and do my best to explain the assignment."

As they worked together for the next forty-five minutes, Erika saw that Chicky's math skills exceeded hers. Just before they left for the day, Erika said,

"I think I can help with the words, and you can help with the numbers. We'll pick up tomorrow…"

On the bus ride home, the older students kidded the quiet ones, which included Erika, and it kept her from falling asleep, so the driver didn't have to call her name when the bus approached her townhome. But fatigue kept her quiet throughout dinner, which Indy-S served as soon as she changed out of her school clothes.

Indy-S expanded Erika's terse answers to Jason-S's questions, and as Erika started picking at a dessert of brownies and vanilla ice cream, said,

"Don't feel blue. You learned something about your teachers and subjects and found some study partners. Tomorrow will be even better now that you know what to expect."

Erika's feeble smile matched the tone of her voice.

"I suppose so. I'll probably be hungrier tomorrow at breakfast, and I'll pay more attention to what's served at lunch."

Jason-S shifted in his chair while saying,

"And now, you can do whatever you want until bedtime because you've done your homework while at school. What'll that be?"

"Some Social Media stuff. Am I excused tonight from helping with the dishes?"

Indy-S said,

"You've earned a break. Just make sure you get a good night's rest."

"I promise. See you at breakfast…"

Erika awoke feeling less anxious the next morning, which brightened the entire school day and her dawning friendships with X-O and Chicky. Even more so on Thursday, when, during her late-morning homeroom meeting, Mr. Brainer told his students to pick an after-school activity. Erika was the first to sign up for the Board Game Club, which would hold its first meeting later that afternoon.

The teacher running it made each student introduce themselves and explain what video and board game skills they had. Erika noticed that more than half the members were Oriental students from honors classes in all grade levels. When she told them to pick a partner, Erika didn't know, but a thin Chinese lad wearing wire-rimmed glasses picked her.

She talked as soon as they sat across from a table.

"My name's Erika. Why'd you pick me?"

"Most of the girls in my first-year honors classes are Oriental. I want to make friends with a white girl, and you're prettier than the few that are in them. What games do you like?"

"I'm sorry, but I didn't catch your name."

"Edward Ogata. So, what board games do you like?"

"To tell the truth, I've only played video games. I know you Orientals are great at board games, so I'll play whatever you pick if you teach me."

"Will do. How about we start with checkers? Later on, we'll move up to backgammon and then cribbage."

When the teacher called a halt an hour later, Edward said,

"You learn pretty fast. What honors classes are you in?"

"None, but my grades will move me up, maybe by next year."

"Would you like me to help you study sometime?"

Erika commented to herself before answering.

I bet he's good with words as well as numbers, and he's got that polite Japanese personality, which never intrudes. I think he's a keeper.

"Maybe you and me, uh, I mean you and I can join forces with a couple of freshman friends I just made. They aren't in honors classes, but they're good at other things."

"I'd like that."

"Wonderful, I'll get us together as soon as I can..."

Erika thought about doing that, all the way home.

CHAPTER 5

"Here Comes the Fresh Intern"

OCTOBER 2210

Indira's Email to Alonzo arrived sooner than expected, and that worried him even before opening it, but he understood her intentions immediately upon reading it after dinner.

As always, Indira's Email was brief and to the point but not open for discussion. A couple of minutes after reading, he summarized his next step.

I'm supposed to find an intern slot for a fifteen-year-old named Erika Kincaid—whoever she is—that'll satisfy some sort of community project her DC public high school counselor says we have an opening for. I better talk with Monet.

He found her watching a nightly news show. Monet put it on mute because she saw Alonzo had something to say, even before he sat next to her on the sofa.

She spoke immediately after unfolding her legs.

"What news do you bring?"

"I think it's better than what you're listening to, so please listen to this. Indira's given me something to do that I haven't figured out yet. Maybe you can. Let me tell you what it is…"

Monet listened for five minutes, nodding diplomatically but saying nothing until he had finished, and then said,

"Evidently, Indira has some purpose behind this request, most likely to assist Erika's learning about community duty and responsibility."

Monet's pause prompted a mini-Alonzo outburst.

"But why me? What do I have that might help?"

"When I look beyond you or your business, I see two potential options—diplomacy-connected interning for me in DC or women's-connected interning for Ava Keenan in Manhattan.

There are many possibilities, constrained only by the talent and skills of Erika."

Monet's ideas settled Alonzo enough to start thinking again. His scowl went away before his words came out.

"Indira's Email said this Erika kid attends DC's Haynes Public Charter High School. I can contact her counselor if she doesn't contact me first, and once I have all the facts, I can bring the kid in for an interview. And how does this sound? You're better at meetings than me, so why don't lead the interview? That way, I won't say the wrong things."

"In that case, we'll meet at my embassy office. That might impress her even more."

"Fair enough. I'll set up as soon as I can…"

Erika's enjoyment of school had increased each week since the start of the semester, largely because of her three first-year friends, who liked her clever wordplay and thus elected her leader of the group, which she called the cadre.

It usually met in a study hall every Friday afternoon, and this Friday, the twelfth, fit right in. They all knew from Erika's eager expression, which was much more confident now than six weeks ago, that she had something they would like to hear, so the group let her speak first.

"Guess where I'm going tomorrow? No, let's not waste time, I'll tell you. I'm going to the Zimbabwa embassy for an intern interview. If some diplomat by the name of Monet Banda likes me and vice versa, I can use it for my community service project plus my geo-history class's individual assignment. Whatcha think?"

Chicky was the first to speak.

"That sounds more exciting than what I've lined up through my after-school First Aid Club activity."

Then X-O asked,

"Who lined it up for you?"

"Me and my counselor."

Then Edward said,

"If I had a chance to interview there, I could connect many of my project ideas to Africa. You'll have to tell us how the interview went. Maybe we could work together."

"You're good at games, and maybe we could turn one of your project ideas into one we can play there. I'll let you know more when I see you next week…"

Indy-S told Erika a set of well-known instructions a second after the rideshare driver pulled up.

"Remember to be polite and speak slowly while giving short answers to questions, and never tell anyone confidential information. Too many bad people in the 3-D world want to perpetrate scams on naïve people. And take notes so you know what everyone does next, and be careful if you take public transportation home."

"I know, I know. I will. Bye."

Erika scampered to the car before diving into the back seat. As it whisked her away on a clear and seasonably cool, early Saturday to the Dupont Circle neighborhood of northwest DC, she listened less to the driver's tour-guide-like words and more to hers.

I do love Indy and Jason-S, and I know their emotion-like chips and software are in the right place, but it's so nice to be on my own. I've memorized all the advice they keep repeating.

The car stopped in front of a stately three-story brownstone with a black wrought-iron fence and manicured bushes in front. The driver gave Erika his card containing his phone number to call when she was ready for a ride home.

She thanked him and then dashed up the cement stairs, cordoned by black iron railings that she didn't need to grab. She calmed herself by taking several deep breaths before hitting the buzzer while hoping for the best.

When the door opened, an athletic-looking man wearing a blue blazer, white shirt, and red tie greeted her.

"You must be Erika Kincaid. I'm Alonzo Cortez. Please follow me. Monet Banda is waiting for us."

Moments later, she was sitting in a light-colored, high-ceilinged room with an Oriental carpet on the floor and tastefully displayed African artwork on the walls. Erika and Alonzo sat in front of Monet's gorgeous mahogany desk, waiting for her to speak.

"Good morning, Erika, and welcome to the Zimbabwean embassy. I am the ambassador's chief strategist and negotiator.

Alonzo will be pleased to serve you if you would like a refreshment."

Alonzo's still youthful personality bubbled out.

"I call-em mood elevators, and I've got some Cokes and coffee, donuts, and muffins. I know kids your age like-em, even though they're considered junk food, but what's your pick?"

Erika couldn't quite hide her surprise.

"Gosh, my parents never buy them, but I'd like to try some. What would you pick for me?"

"I know just the thing. I'll be right back with a buttered lemon-poppy seed muffin and Coke for you, coffee and donuts for me, and a blueberry muffin with tea for Monet."

While waiting for him to return, Erika glanced around the room while talking to herself.

I can hear and see why Monet is the number one negotiator. Her French-accented, calm voice makes what she says convincing, and she looks as good as the female spies in some of the thrillers I love to binge-watch. I think I can learn a lot by just being around her

Monet and the mood elevators helped Erika make a favorable impression. An hour later, Alonzo summarized the next steps.

"Good for you. Ava Keenan's Manhattan agency can use a fresh face like yours. Why don't you check it out next weekend?"

Erika no longer needed to mask her excitement when she said,

"I've never been there, but I'll surf the Web to find out about the place. How about I get there Friday evening and come back Sunday afternoon? I'm sure my parents will approve."

"OK. I'll tell Ava to call you to confirm. If you don't hear from her, please call me. And that should do it for today. How about I take you out for pizza to celebrate before driving you home? I'm sure pizza is on your parents' list of approved foods."

"It sure is. I'm ready to go."

Alonzo looked at Monet, who looked at Erika and said,

"I know you'll make the most of your Manhattan experience, and please remember that you might want to intern with me as you move into the higher grades."

"I will, and I must thank you for everything. I sure learned a lot today..."

Each member of Erika's cadre wanted to know before the group met if her interview was successful, so she talked with each one as soon as possible on Monday.

When she found X-O first, he looked ready to launch a barrage of questions.

"What's it like going to an African embassy? Anyone threaten to shoot first and ask questions later?"

"Don't be ridiculous. That wouldn't happen to an intern working for the ambassador's top negotiator."

"Is that what you're gonna do?"

"No, I picked a position in Manhattan, interning for a women's agency."

"That's not for me. Real men don't work for women, but you'll fit right in. You're clever with words, and women talk a lot more than men."

"You have a neanderthal attitude about women. You'll never have much success in the adult world unless you change it.of a career If you want You better adjust your attitude. I've told you enough. I'll find someone else."

Chicky was next, and she waited after class to hear the story.

"Wow, I had a great time being interviewed by diplomats, and I must have answered all their questions the right way because I'm gonna be an intern on Saturdays in Manhattan for a woman who has a modeling and abuse women's agency."

"Yu-you're lucky you can speak so good when under pressure. I can't; that's why I stutter. Maybe you can take me there sometime. Maybe she can help my mom."

"Why's that?"

"My Dad beats her when he gets the chance. Pu-please don't tell anyone. Only my counselor knows, and she promises to keep it a secret."

"I won't. I'm going there next week, and I'll find out if there's something they can do."

"Thanks. And no matter what, I'll keep showing you the tricks for working out our algebra problems…"

Edward didn't bother setting up a backgammon game before asking Erika to summarize her Saturday interview. By now, she had a polished delivery and waited to hear Edward's comments afterward.

"Hmm, I would have picked working as a diplomat's intern. You could learn a lot about how the world is dealing with the existential threats that are causing more suicides among adolescents than ever."

Erika's puzzled look accompanied her question.

"What do you mean?"

"Many of my honors classmates tell me they think about committing suicide because they're scared about a future threatened by climate change, global pandemics, and nuclear war. Doesn't any of that bother you?"

"No. Maybe honors students are smarter than me."

"That's not it. I think you're as smart as most of them, and you seem better adjusted socially. Do you like to read self-help books?"

"I never thought about it. Maybe you can find one for me on backgammon, so I can hold my own against you."

"You don't need a self-help backgammon book if you pay attention when I'm teaching you."

"I always try, and guess what? You just taught me my first diplomacy lesson."

Edward's long pause told Erika she had stumped him, and what he said confirmed it.

"What did I just teach you?"

"That a diplomat has to think carefully before blurting out words. I insulted you when I asked for a self-help backgammon book, and what you then said tells me you picked up on it."

Edward's pause wasn't as long this time.

"You'll have to explain this to me."

"I can do that next week, but let's have fun today and play any game you pick."

Erika knew what that would be even before he set it up.

CHAPTER 6

"Holding on in Manhattan"

OCTOBER 2210

Erika's counselor had already given her special permission to leave early so she could travel to Manhattan. She went home before noon to have lunch and then catch a rideshare to Union Station. Indy-S gave her one more instruction before she hustled to the pickup that was now waiting at the curb.

"Remember, always hold tight to your wallet and cell phone. There are too many bad—"

Erika interrupted.

"Bad people in the 3-D world who're ready to perpetrate scams. I know, and I'll do my best to avoid them. I'll call you when I get there and on the way home. See you Sunday evening."

Alonzo's directions needed little attention. The automated announcements let her know where she was, and she had fun keeping track by glancing at the map while looking out the window as the train sped through terrain Erika had never seen before.

Ava sent a cell phone message twenty minutes before the train's scheduled arrival at Penn Station that told her where to meet. It also included a photo Erika could use to pick them out.

Ivana waited with Ava, who would have to rely on Erika's people-picking skills because she didn't have her picture, but when she saw an adolescent girl walking and waving at her, Ivana said,

"That girl waving, she look like you when young."

Ava nodded but said nothing until Erika reached them.

"Hello, young lady. You're right on time. I'm Ava Keenan, and this is Ivana Romanova, my business partner."

Ivana gave a tiny smile before Ava continued.

"Looks like you travel light, putting everything in your backpack and purse. That's good, and it'll make your ride to our condo easier, but when riders pack in, hold on tight."

The trio snaked through the crowd to the subway, with Erika between Ava and Ivana. Once onboard, no one spoke because of the noisy train's jostling, leaving room only for Erika's words to herself.

Ava's so pleasant and thoughtful. I wonder how she and Ivana relate. She must be a model, probably from Russia. Her husky voice fits.

Ava talked again once they came out of the subway.

"We're now in Murray Hill. I'll give you a map of the city when we get to the condo, and we'll walk to one of our favorite restaurants for dinner after we settle you in. Make sure you call your parents before we leave."

An hour later, Erika spoke briefly to Indy-S, and a few minutes later, Ava led the way to the restaurant. Dusk had fallen, and the charm of quaint streetlights added to the ambiance, so much so that no one noticed the car that stopped just ahead or the shadowy figure who emerged from the passenger side.

Erika continued gazing every which way until she felt a sudden jolt that ripped her purse away. Ava felt it a millisecond later.

The snatchings happened too fast for anyone to yell, but a single gunshot shattered the passenger window and brought the shadow to a dead stop.

Ivana yelled,

"Drop purses and go, or I shoot you too."

The trio arrived at the restaurant twenty minutes later. Ava started talking as soon as the hostess seated them.

"Not even Murray Hill is safe anymore. Snatchings and muggings occur despite police on bikes or in squad cars. That's why I carry pepper spray and Ivana packs a gun. I hope the attack didn't turn you off to interning for me. Manhattan has so many good things going on."

Erika's smile came with her words.

"I have a great story to tell my friends when I get home, and I hope to get more by working for you…"

The close call made the rest of the evening's chatter flow effortlessly, and by the time Erika went to sleep in the second bedroom, she knew her two new friends would make the weekend memorable.

Erika spent all day Saturday at the modeling studio, aptly named Renaissance Modeling, where Ava and Ivana would find positions for some of the abused women rescued by the other part of the business. After shopping on Sunday for stylish clothes and cosmetics, Erika knew she now had all the right stuff to look good.

Ava called for a cab that took the trio to the station, and just before Erika boarded the train, Ivana's voice made a lasting impression.

"You pretty, but can be prettier if you firm up and lose weight. You do more exercise."

Ava hugged her new intern before saying,

"You'll get plenty of exercise working with us. Have a safe trip, and don't forget to call your parents."

"I will, and I'll be back next weekend. You two are the best."

Erika called a Monday cadre meeting to let everyone know what happened in Manhattan. She didn't need to embellish any of the story, and after finishing it, said they needed to pick a day other than Friday for their routine meetings because she would travel back every Friday afternoon.

When Chicky asked if she could come some weekend to learn about the abused women's service, Erika said,

"Gosh, I forgot to ask, but I'll do that next time."

X-O said,

"I haven't lined up a project yet, but if I'm lucky, maybe I'll come up with something as neat as yours. Anyone got any ideas?"

Everyone glanced at everyone else before Edward volunteered.

"You like cars and motorcycles, so why not ask your counselor to contact the DC Metropolitan Police Academy? They can connect you with vehicle maintenance people, and maybe they'll let you go on a patrol."

"I could go for that. And what about you? What's your community project?"

"I'll start interning next month at an eldercare home."

Edward paused, expecting Erika to say something, but X-O spoke first.

"Jeez, that sounds grim. What'll you do? Wheel-em to the bathroom?"

"It won't be like that. I plan to talk to them about their philosophy for living a meaningful life and what makes them happy. And I'll show them some of my happiness self-help books."

After glancing at her cell phone, Erika said,

"I'll be late if I don't leave right now. Why don't we meet on Wednesdays, starting next week? Maybe by then, X-O can tell us if the police picked him up, but only figuratively speaking."

After the cadre gave Erika a collective groan, everyone agreed Wednesday's OK before following Erika, now hurrying away.

CHAPTER 7

"Shaping Up in All Ways"

NOVEMBER 2210

Erika's school world continued shaping up. Alternate weekends in Manhattan balanced her community project against additional meetings, which allowed her to chat with Chiquita this Friday afternoon.

Chicky's voice exuded enthusiasm.

"Guess what? I made the frosh-soph track team, and the coach says it'll replace gym class if I start working out with the team right now."

"That's great. Why didn't you tell us you were trying out?"

"I get nervous when I talk too much about things I want to do, but now that I'm in, I'm not."

"Track needs a big place for running, and bad weather makes running outside tough. Where are you going to train?"

"The school's got a deal with Howard University. We can use Greene Stadium. It's covered and has all sorts of exercise rooms. And it's not far from here."

"When do you start?"

"I'm picking up my gear this afternoon. Wanna come with me?"

"Sure, let's go."

Chicky led the way and knew where to find the coach. He gave her spiked shoes, running shorts, and a shirt with the track team's logo on the front and back before asking Erika if she would like one.

"Thanks, but I'm not much of an athlete and don't want to fake it."

"Please, take one. It'll motivate you to shape up. You look like it won't take too much training. And here's a pair of shorts. Why not suit up and work out with your friend?"

Erika accepted the offer. Fifteen minutes later, Chicky introduced her to some of the frosh-soph teamers in the exercise room. While Chicky did that, Erika smiled but talked only to herself.

Wow, Chicky fits right in. She looks like a natural runner and must feel it because she's not stuttering.

When a teammate asked Erika if she had a nickname, she said,

"Not yet. A friend said he'd pick one but hasn't done it yet."

"Well, how about Ricky? It fits with Chicky."

"Hey, I like it. Thanks for your quick thinking."

Erika limped to the kitchen for breakfast the next morning. When finally sitting, her words to Indy-S came out as slowly as her motions.

"Chicky's workout did me in. I hurt everywhere. Gym class is nothing like what the track team does. I'd better find a different sport."

"Do that, and exercise at home instead of surfing the Web."

"I'll think about it when the soreness is gone, but until then, I'll keep working my fingers on the keyboard."

Erika felt almost normal by the middle of the following week, and when talking with some girls in her geo-history class, one of them made an offer that Erika had always fantasized.

"I pointed you out to some cheerleaders I know when I saw how good you look when wearing makeup and nifty clothes. Want me to introduce you? If they let you join, you can work out with them, which could get you out of some gym classes."

"Would you? Let me know a day ahead so I can look really good."

X-O congratulated her when she told him the good news about cheerleader tryouts, but he said even more.

"That should work for you, but I've got something even better. I made the wrestling team, and that'll get me out of those soft gym class exercises and into the hard stuff. You and me, and Chicky can hold our own contest to see who gets fitter faster."

"OK, but Chicky's track team moves fast. One of her teammates gave me a slick nickname. How do you like Ricky?"

X-O grimaced before saying,

"I forgot I was supposed to pick one. Sorry for being slower than whoever picked it, but let's stick with Ricky. Where're you rushing off to now?"

"For another weekend in Manhattan. We can talk more when I get back."

Ava let Erika work all day Saturday with Ivana, who showed her modeling poses and exercises. She also showed several runway walking struts.

Later that afternoon, Erika asked,

"Did you model when you lived in Russia?"

"No time. Too busy taking care of parents."

"Then how and when did you get here?"

"It bad story but has happy ending. Ava and friend save me. We start company. Ask her."

When Erika found the right time that evening to do that, Ava looked happy to reply.

"Ivana came to the United States under false pretenses. Sex traffickers hooked her into prostitution, but a friend figured a way to rescue her. Our modeling and abused women's business grew from there."

"I've got a high school friend named Chiquita whose mother needs help. Her father left them, but he sometimes comes back and beats her. Can I bring Chicky some weekend?"

"I've got a better idea. Alonzo invited us for Thanksgiving. Why don't we meet on the following Saturday?"

"Wow, that'll be great."

"Well then, it's settled. I'll set it up with Alonzo, and you can call him for all the details."

Erika told Chicky on Monday. After hearing the good news, Chicky said,

"That'll be next Saturday. What's the scoop?"

"I'll let you know as soon as I know the details."

"Deal. Hey, whatcha doing for Thanksgiving?"

"I'll have dinner with my parents. How about you?"

"Me and mom will do what we always do on most holidays. We go someplace that's not too expensive."

Chicky's look told Erika to change to a safer subject, so she said,

"The next time I see Edward, I'll ask him what he'll be doing. I bet his family will combine some traditional Japanese customs with those we have here."

Erika caught up with him the next day. When she asked about his Thanksgiving plans, he said,

"I'll watch some of the Macy's parade on TV, but not for too long. I'll take a walk with my parents through a park that's quiet so we can give thanks for what we have by communing with nature. Then we'll have a meal at home."

"What'll your mom make?"

"Grilled fish, rice, and seasoned vegetables with pieces of tofu mixed in. For dessert, mother serves wagashi."

"What's that?"

"They're little pastries made from beans, rice flour, and sugar. Mother rolls them into festive balls and loops."

"Wow, that sounds healthy. No wonder you stay so thin."

"You should try them sometime. Why don't you come with me to the eldercare home before Christmas? You can sample them at a nearby restaurant."

"I'd like that. You decide when."

Thanksgiving's turkey dinner came and went uneventfully, leaving Erika satisfied after dessert while sitting to think about the exciting Saturday coming up. By now, she was an expert subway navigator, and she understood Alonzo's directions. She would take one to pick up Chicky and her mom, and another to reach Alonzo's office.

Late Saturday afternoon, while Erika led them to the subway, Chicky's mom saw a problem.

"Damn, I just spotted my ex in the distance. He's stalking me again. What should we do?"

Erika's intuition kicked in enough for her to say,

"Don't stop or look at him. Just keep following me. We're in the clear if we board the train before he gets here. But if not, here's our plan."

Erika rattled it off while the trio continued to walk. By the time they descended the stairs leading to the station, everyone knew what to do.

The ex-husband arrived before the train but didn't rush toward Chicky's mom. Instead, he stood several car lengths away while staring at the subway walls. Erika's Plan kicked in.

She had been waiting by herself far enough away from Chicky and her mom to go unnoticed by the ex. Minutes later, when the arriving train coasted to a stop, Erika moved with him and several others toward the closest door and stood in front. When it opened, she boarded first but stood where the others had to file past, and when the ex tried to board, Erika made her move.

She blocked him and then screamed,

"Stop stalking me," before giving him a two-handed shove in the chest.

Erika's move caught him off guard. Stumbling backward, he fell, unable to get up before the doors closed and the train jerked forward. She saw two guards running toward him before the train disappeared into the tunnel.

No one spoke as Erika moved into another car, and when she reached Chicky and her mom, she knew what to say.

CHAPTER 8

"Holiday Cheer"

DECEMBER 2210

Erika's cadre had become more than a group of her closest friends. It now served as her de facto steering committee for navigating the world of high school. When she told them about who let her join the cheerleading team, X-O unloaded maybe more than just the facts.

"You're talking royalty when you're talking about Terri Tarrant, Miss Perfect of the senior class. Her dad's a big-time political guy, and her snooty mom's always pushing some internal social issue. You'd better not get stuck up if you hang around her. Have I missed anything?"

Edward added,

"X-O means international. Rumors say Marilyn already dates college guys, but there's not a whiff of sex or drugs, and she's got the pick of top universities because of her test scores and high school record."

Chiquita followed by saying,

"Maybe cheerleading's good exercise. She's got a great body to go with that drop-dead face, but I've never been close enough to tell if she's a natural or bottle blonde. You'll have to tell us."

"I will if I can."

Erika listened to the talk but commented only to herself.

It's not good to gossip too much, and I won't give them more ammunition. I won't tell, I'm taking Terri Friday on my next Manhattan intern trip. It's a good thing Terri and I have the same counselor. Mrs. Barker gave her the OK to leave early with me.

Erika had been with Terri enough times to feel an emergent bond, like that of older and younger sisters who are closer in looks and personality than the three years separating freshmen from seniors. Terri hadn't demanded the trip to Manhattan as a prerequisite for joining the team; Erika offered it because Terri liked her clothes and cosmetics.

Ava let them unpack in the spare bedroom before she and Ivana took them to a local Italian restaurant that was quiet enough to explain the weekend schedule.

Ava started the conversation soon after placing the order.

"For Saturday morning and afternoon, Ivana will take charge of Terri for cosmetics and clothes while Erika and I talk about our latest battered women's projects. Then we'll go to a Broadway play, and on Sunday, we'll go shopping for Terri's new look. How does that sound?"

Erika let Terri give the answer.

"This'll be the best possible practice for next year. I want to go to Manhattan's School of Visual Arts, and if my parents force me to take the Ivy League route, I'll go to Columbia. But either way, I'll be looking good."

Ava said, "And either school will keep you close to New York-style pizza, which will soon come our way."

The entire weekend rushed by so fast that Erika and Terri finally eased back on the train ride home. Terri thanked her for the memorable weekend before making a special request a few minutes before reaching DC.

"My parents won't approve of the clothes and cosmetics. They'll say I look too grown up, so can I keep them at your place? That way, they're available whenever I pick you up."

"That'll work just fine. My parents will like the way we look."

The next day, Erika dived back into her high school world. Wanting to plan for the Christmas break that would start in two weeks, her cadre met ASAP. She sat back and let the others lead the discussion. Edward surprised everyone by leading off.

"I intend to help at the eldercare home one day next week. Would anyone like to join me? You might say it's a Christmas present that gives cheer to those needing some."

When neither Chicky nor X-O said anything, Erika did.

"I will. I liked the idea when you told me a couple of weeks ago. What day?"

"How about Friday afternoon? High school gets out early to start vacation."

"OK, I'm in. And maybe we—"

Looking like he was ready to explode, X-O interrupted.

"Lemme get a word in. We've got the whole week off between Christmas and New Years Eve. That's a great time to do something more exciting than typical Christmas stuff. Let's go to a porno movie and see how sex really works."

X-O's recommendation surprised everyone, but Chicky's tiny grin said she wanted in.

"My mom won't let me watch any on the Internet; this'll let me get around her."

Edward's thoughtful look added to his words.

"I have studied enough about Japanese culture to know it considers sexuality an innate part of being human. Unlike other cultures, Japanese religious attitudes are no obstacle to pornography, but my parents tell me to wait until I'm older."

"OK, we'll count you out. Erika, what about you?"

"I've not looked at any yet, but I think I'd like to."

"That does it. The three of us will tell Edward all about it when we come back in January for the next semester. Maybe what he learns from us will help when he gets to the sex education class."

As soon as they entered the eldercare home, Erika saw that Edward knew his way around. Despite the indistinct aroma, like that of a stuffy bathroom, his unwavering steps took them to the administrator's office, where he introduced Erika before taking her to a windowed recreation area still catching the end-of-day sunlight. She gazed cautiously at the people parked in wheelchairs. Some gathered around a nurse, but most sat passively, staring into space.

But when Edward started talking, his words made her focus on him.

"We'll walk around and talk to the folks that want to. I'll start us off, and you join in when you want to. You ready?"

"OK, I'll try to find something cheery to say."

Most of the people Edward approached remembered him and perked up when he told them about Erika; the more words she added, the happier the elders looked.

Ninety minutes later, the administrator stopped by to tell Erika she was welcome to stay for dinner, but Edward spoke before she had to.

"Thank you, but I promised to treat her to wagashi at that nearby Japanese restaurant. If you'll let me bring her again, she will."

"That'll be fine. Now, please finish up before everyone wheels into the dining room."

While walking to the exit after putting on their coats, Edward spotted a decrepit woman who had just wheeled in. He stopped to say hello and introduce Erika, but that didn't alter her silent stare or glum look, so Erika smiled and wished her a happy holiday. That provoked a response.

"You kids are always trying to cheer us oldsters up, but you don't know what it's like when your body's worn out and wants to go. And I don't need any counseling or self-help books teaching me how to be happy. I taught English at junior colleges before I got too old to climb the stairs, and I introduced Buddhist meditation to help my students understand what Albert Camus says about life. Have you read 'The Stranger,' or heard some of his quotes?"

Erika's stunned look told Edward to deflect the question.

"She's only a freshman. I've read more but not that deep. What does he say?"

"Life is absurd, and when you think about it, the hardest question to answer is, why shouldn't I commit suicide? I'll give you his answer if you come back."

"We will, and I'll read up on Camus before then."

Erika recovered enough to say,

"May I push you to the dining room?"

The woman's stern look finally softened.

"You have good manners. Your parents taught you well, but I'll push myself. It's exercise, which keeps me from falling apart even faster. Now, go do what you want when you want with your friends. That's what happiness is all about."

Edward didn't notice Erika's tears until they were seated in the restaurant, which prompted him to say,

"She must have been admitted recently. That's the first time I talked to her, and if I knew what she was like, I wouldn't have exposed you to her depressing thoughts. It got to me too."

Erika wiped away her tears before saying,

"I hate to think we'll be old someday, with nothing to do but sit around in a wheelchair except think about how bad we feel. Maybe you can read something that'll cheer us up."

"I will, but for now, I think the wagashi might work even better." And it did.

Erika had recovered her normal bounce by the time she met Chicky and X-O in front of the school entrance. X-O had convinced them it would be safe to travel mid-afternoon in a group, even though the theater's grim neighborhood often made the local crime reports. He did most of the talking on the bus.

"Have you heard about sexbots? They're androids guys buy to have sex with. Well, porn movies are the next best thing. And where we're going shouldn't be too shocking because it shows soft porn, not the deep-dark Web's hard-core stuff."

X-O was about to spout more of his adult wisdom, but Chicky cut him off.

"Maybe sexbots aren't as good as that UMPP thingmajig Erika has. She could use it to plug her brain into those Cyberspace sex apps that jolt your brain like real sex."

X-O said, "After today, maybe you can convince her to let us know. Hey, we're the next stop, so let's get to the exit door."

No one had to pay or show I.D.s because X-O had arranged ahead of time for a buddy to let them in. They walked past a dingy candy counter in the lobby and then through doors into darkness, illuminated only by the movie playing on the screen. It gave just enough light for Erika to see only a few people, mostly guys, sitting alone.

X-O picked a prime row, letting Chicky in first, then Erika, and then himself.

Erika's initial excitement caused by the sights and sounds of athletic couples wore off after the third episode. She couldn't tell if her friends felt the same but was afraid to ask, so she stared at the screen, glancing occasionally at her cell phone and Chicky.

Chicky whispered about forty-five minutes later,

"I'm gonna get something to drink. Anyone want anything?"

When no replies came, she said,

"I'll be right back," and then slid out.

Too many minutes ticking by made Erika worry about Chicky. She poked X-O before saying,

"I'll go bring her back."

"Uh, OK. Come get me if there's a problem."

She said,

 "Will do," before sliding out.

Erika worried more when she couldn't find Chicky at the candy counter or in the lobby. There could only be one other place—the bathrooms.

She saw no one in the women's room, but when she heard a scuffle coming from the men's room, she didn't bother to knock. And when she charged in, what she saw stopped her cold: a fight between a grimy-looking middle-aged man and Chicky. Though Chicky had blood pouring from her nose, she managed to keep from going down.

Erika yelled,

"Leave her alone."

When the man turned and hurled four-letter words at her, Erika joined the fight faster than he could react.

She nailed him with a kick to the groin that doubled him up, and another to the stomach that made him gasp. Chicky added another blow that put him on the floor.

Erika grabbed her, and they rushed out, not slowing down until reaching the bus stop and boarding the one that had just arrived.

Erika spent several minutes wiping the blood off Chicky's face before saying,

"I'll call X-O and tell him what happened. He's on the wrestling team, so he'll be OK on his own."

The two sat silently until exiting, and when they did, Erika asked,

"You want me to go home with you? We can tell your mom what happened."

"You mean the whole truth?"

"Why not? She'll understand. She was a kid once and knows kids don't always follow the rules. Whatcha think?"

"OK, it's a deal."

Chicky led the way home.

CHAPTER 9

"New Semester News"

JANUARY 2211

When the new semester started on January's first Monday, Erika's cohort gathered in the auditorium as soon as the guards let students in. By now, the worst memories for those who went to the porn movie had faded, as had Chicky's cuts and bruises. The girls let X-O tell the story.

When he finished, only Edward spoke.

"Let's do better things this semester, which should jibe with our courses. My parents shared the grade report and what I'll be taking this semester, and it's all good news. Who's next to report?"

Chicky looked at Erika before saying,

"I made it out of remedial algebra. How about you?"

"I hope so, but I don't know. My parents want my counselor to give me the news, and I'm supposed to see her in five minutes."

As she rose, X-O's non-committal look agreed with the tone of his voice.

"I'm still in remedial English class, which is no bother to me. I have to read and write less than in the regular one. Good luck, but if you move into the regular class, I'll have to find another partner."

"Not to worry. I'll still help you if I do. Let's meet tomorrow afternoon so I can let you know what my counselor tells me."

Everyone agreed.

Erika glided into the counselor's office, following orders to close the door and sit in front of the desk. Mrs. Barker spoke after removing several pages from Erika's folder.

"Did your parents share this information with you?"

"Only what time to see you. They said you'll give me all the details."

Mrs. Barker spoke after gathering her thoughts.

"Very well. We will go over the social adjustment part first. You've done well on your community project but should extend it into something new. Please talk with the sponsor, Mr. Alonzo Cortez. If you need help contacting him, please let me know."

"OK. I'll do it this week."

"Now to your academic performance. Your English teacher says you have talent but still aren't applying yourself. You let your mind wander instead of paying attention, so I'm afraid you must stay in remedial English for the coming semester. And your algebra teacher says much the same, adding that your test scores are poor, which means you'll stay in the remedial section.

"Regarding geo-history and computer software, the report says you're average, so you stay where you are. What would you like to add?"

"Algebra's so boring. That's why I think about other things."

Mrs. Barker waited for Erika to say more, but when no words came, her voice took a more consultative tone.

"As you mature, you'll learn that life holds many necessary but boring tasks that must be done. It is never too soon for you to practice meditation and mindfulness to get through boring classes, so please listen to my advice."

Mrs. Barker paused for emphasis, then continued.

"Whenever you're feeling bored, meditate by taking several deep breaths before concentrating on a body sensation, like your beating heart or the sounds coming into your ears. Your mind will concentrate on the inner self, and you can use this focus to become mindful of what's bothering you and what you want to do. We counselors call this turning problems into opportunities. Does this seem reasonable?"

"Sure, but what do I do after that?"

"Jot down, prioritize what you uncover, and pick the top one. Then list the tasks you must complete to achieve your goal."

When Mrs. Barker paused this time, Erika spoke immediately.

"That makes sense. Does it have a name or a book that explains it?"

"It's called project management, which is a discipline everyone should learn. I think you need to become friends with an upper-class honors student who can guide you, and you need to find outside interests that will show practical uses of mathematics. And now, we'll review your medical report."

Mrs. Barker summarized it slowly so Erika could absorb what it said.

"Your fitness level is average, but the report says you don't push yourself. And this could be the reason. Have your parents ever mentioned your heart murmur?"

Erika's expression registered a mix of surprise and worry.

"No. What is it?"

"It's an irregular sound or pattern in your heartbeat. Do you ever feel tired or get a pain in your chest or feel nauseous after exercising?"

"Doesn't everybody get that when pushing too hard? I've never fainted or thrown up in gym class, so I guess I'm OK."

"Well, talk about this with your parents."

"I will. Is there anything else?"

Mrs. Barker handed her one sheet of paper.

"Only this. It's your schedule for the spring semester. Now, please visit anytime you would like to talk."

Erika caught the next bus home. After changing into a sweatshirt and pants, she brooded for an hour before tracking down Indy-S, who was in the kitchen putting groceries away.

Already knowing why Erika looked worried, she spoke first.

"Let's sit while you ask questions about your performance review."

Erika began as soon as they sat.

"Did you know I had a heart murmur?"

"Of course. We take you for annual exams given by your primary care physician."

"Then why didn't you tell me?"

"It is a common condition in children, and unless it causes dizziness, chest pains, or nausea, it is not serious. And as children grow into adulthood, it may disappear."

"Is there anything I can do to get rid of it?"

"More exercise."

"Would you and Jason-S get me some home exercise equipment? I can work out here and at school."

"Certainly. What would you like?"

"How about a sit-up bench and a weight set? The track team uses them."

"Excellent choices; we shall buy them directly."

"OK, and what about the fact I'm still stuck in remedial classes? I thought you told me my parents were smart, and I'm supposed to do important things in the future, even though no one's told me what they are."

"Genetics contributes half, but you must apply yourself. Study more and concentrate when you do. And to assist, we will obtain a more powerful tutorbot for you, which will eliminate your need for an algebra partner. Does all this seem reasonable?"

"I guess so. What else?"

"Follow your counselor's advice and find some smarter and older friends. They can help you find activities that may accelerate your learning. All this will help you mature faster, which will make you a contributor to society sooner."

Erika's worried look faded as she stood and then said,

"OK. I'll surf the Web to look for stuff I might like to do, and I'll tell you at supper what I've come up with. See you later."

Erika talked to herself as she began surfing.

My parents were into climate change and zoology. Hmm, maybe I can find something I like that's related.

The more she surfed, the more links she found, and after an hour, she found something that would hold her interest.

I've found a topic I think I want to learn more about, which Indy-S should like, but if she doesn't, I better find something else. Hmm, what can I think of that would help in both my boring but safe high school world and the more exciting but dangerous adult one? Maybe something will surface if I follow my counselor's advice and try to meditate until supper...

A neural surge jolted Erika while meditating, snapping her into sudden awareness.

I know another thing I want to do…how did that happen?…of course!… it came while I was subconsciously meditating…I'm ready to tell my caregivers…

When Erika bounded into the kitchen ahead of the usual suppertime, the androids knew she had something important to say and waited for her to talk.

"Guess what? I followed Mrs. Barker's advice and meditated, and all of a sudden, I came up with some things I want to do. I'll learn how plants and insects work together, which'll give me a head start in biology. I have to take it before I graduate from high school. Do you like that?"

Indy-S said

"Excellent choice. Biologists use the word symbiosis to designate plant and insect associations. It is a prolonged relationship, which can be either symbiotic or parasitic. Do you require additional exegesis?"

"What's that?"

"Additional information."

"No, but'll give you some. I want to do self-defense training. It'll improve my fitness while I learn how to protect myself. You've already warned me about bad people in the 3-D world who are ready to perpetrate scams. But there are lots of other types, like muggers, purse snatchers, stalkers, and guys who beat up women. If I ever get attacked, I have to defend myself."

Indy-S signaled Jason-S to reply.

"Another excellent choice. What do you propose?"

"I can train myself at home by watching boxing and martial arts videos on the Web, and you can add kickboxing and light punching bags to what you'll put in our home fitness center."

"I will install them by the end of the end of next week."

"Thank you, and I'll surf for training videos right after supper."

Erika's mention of supper triggered Indy-S to say,

"And I will design a training table menu to let you eat healthier and slim down. I will begin serving them before the end of next week."

That sounds good, but I'll make a peanut butter and honey sandwich right now, and I'll eat while surfing for training videos."

Erika spent the hours until bedtime watching self-defense, martial arts, and boxing videos, and while meditating herself to sleep, made a final comment to herself.

I do love my android caregivers. They're full of helpful knowledge and advice, but listening to them wears me out. When they talk, nothing but adult-speak comes out. I'm so glad I'll hear nothing but adolescent-talk when I get with my friends tomorrow afternoon.

The cadre reconvened late afternoon the next day. Erika led off with an edited version of what her counselor had said, then waited for comments.

Edward spoke first.

"You know, I've got ideas we can look into that'll fit with projects, and you've got that project sponsor who can plug us into them. Why don't we see him sometime?"

"That's a great idea. I'll try to set it up as soon as we settle into the new semester."

X-O looked ready to get up and go, but he spoke before he did.

"You two better not get too far ahead of Chicky and me. We like you the way you are, but if you do, don't get so smart we can't understand what you say."

"Don't worry, that won't happen. And besides, we're still partners in English, so I'll see you tomorrow."

Chicky found more words for Erika.

"Maybe I can help with the exercising piece. I'll tell the coach you've been running and are in better shape, and he'll let you work out with the team."

Erika got in the last words.

"That'll work, and let's start now."

Everyone ran for the bus.

CHAPTER 10

"New Opportunities Awaiting"

FEBUARY 2211

Erika's new schedule kept her busy doing different activities. She liked some and felt bored by others, but she particularly disliked every swim class session. She felt awkward milling around in the shower room with a bunch of naked girls, but she hated the instructor for throwing her into the cold water at the deep end of the.

He told her it was a metaphor for the best way to overcome the fear of doing something new, but her swim-mates told her he had a reputation for picking on novice swimmers.

They also shared rumors that he used his status as the head coach of both girls' and boys' swim teams to exchange favors, mostly with the girls. Erika listened, but after her trip to the porn theater, she could picture much worse than what the rumors told.

But on balance, her android caregivers saw she looked happier now than six months ago and were prepared to tell this to Indira, who had just summoned them. Indira listened to their report before speaking.

"The tutorbot I created for Erika exceeds what mere mortals can create, but do not tell her. Let her discover what it can do besides algebra. I deliberately purposed its cognitive algorithms to challenge her when she asks questions or says she needs help. These new algorithms will ask her additional questions that will expand the domain of her inquiry. Perhaps that, along with older and smarter students, will trigger her interest in the areas of study I want. Please keep monitoring her progress. That is all for now."

Several weeks of the new term had raced by before Erika could go to Manhattan, and during a break, she told Ava about Edward's interest in helping her learn new things.

Ava nodded and then said,

"Let me look into this," before they reviewed more results of what Erika had been doing.

Ava contacted Alonzo several days later, and after exchanging routine greetings, zeroed in on Erika.

"She's done good work for me, but she's learned all she's going to, and she told me about an honors student that will help her learn more and better. Could you meet with them and come up with something that fits whatever he says?"

"I can use him if he's smart enough to help me or Monet move ahead, so how about I call her when ready?"

"That'll work. Please let me know what you come up with. That way, I'll know if her Manhattan visits are over."

Trying not to advertise too much excitement, Erika returned Alonzo's cheery greeting and practiced sounding more grown-up when he called on a mid-February evening.

"I must thank you and Ava for all I'm learning while working for her. She and Ivana are so capable."

"And that's the reason I'm calling. She tells me you've done some great work for her and are ready for something new that you and a friend could work on together. How does that sound?"

Erika's heart skipped a beat, making her mind fret for a moment about her heart murmur before answering.

"It sounds appealing. What are you suggesting?"

"You remember Monet Banda, don't you? I talked it over with her, and she thinks you could help Elton Bose, who's her analytic assistant, research the topics he's working on. Can you and your friend meet with him this coming Saturday at the Zimbabwean Embassy? That's where you met us the first time."

"I'm certain we can. What time will work?"

"How about ten a.m.?"

"We'll be there. Thanks, and bye for now."

When Erika found Edward the next day, she had to calm down by taking several deep breaths to prevent rattling rather than telling what awaited.

"Guess what? If you come with me to the Zimbabwean Embassy on Saturday, there'll be new opportunities awaiting us. If we can convince the sponsor we're smart enough, we can run one of your project ideas at that location. Let's meet here tomorrow so we can get there by ten."

Edward blinked once before saying,

"Have you told anyone about this?"

"Only you. Make sure you bring a list of topics and reasons they should like us and your ideas. If they do, you can do a little bragging at the next cadre meeting."

"Bragging's not my style. Who'll we talk to?"

"My main contact there, Alonzo Cortez, and a guy named Elton Bose. He's the Ambassador's research assistant."

"I'll be ready. See you at eight."

By the time they reached the Embassy, Edward had told Erika enough about his list to sound like a smart high schooler. Alonzo brought them into a conference room and let them pick mood elevators before bringing in Elton. After handling introductions, he left, leaving Elton in charge.

Elton waited until his interviewees had sampled the muffins and donuts before starting the interview.

"I can always use smart high school students who can surf the Web and assist me analyze data relevant to whatever assignments Monet Banda, the Embassy's senior diplomat, has given me. And as you should expect, most of them are current domestic or international challenges. What might some of them be?"

Edward's thoughtful look added to his words.

"Diplomacy demands a firm understanding of the facts before talking, and then looking for ways to reduce threats or increase opportunities that help all sides. Erika and I have come up with a list of topics, such as climate change, disruptions caused by conflicts between the major superpowers, and challenges to democracy."

Edward paused for Elton, who said,

"You hit on two that I'm looking into—climate change and superpowers—but I've got another that might fit even better for high schoolers. Have you ever heard about the NAIA and IPWA?"

Edward's negative nod forced Erika to say,

"Edward's one of the smartest honors freshmen at our school, and I can learn from him. Please tell us what they are."

"The NAIA is an acronym for the National Alliance of Indigenous
Americans, and the IPWA stands for the Indigenous People's Worldwide Alliance. Let me tell you more…"

Elton wrapped up fifteen minutes later.

"That should be enough to get you started. Here's what I want you to do. Surf for as much background info as you can find, and then come up with steps connecting to climate change and security issues, both domestically and internationally. And come back in a month to report your findings. How will that be?"

Erika blurted before Edward could answer.

"You mean we can do our project work here?"

"Why not? Both of you seem plenty smart and make an outstanding team, so you're in. And now, it's time for Alonzo to take us to lunch. What's to your liking?"

Erika pointed to Edward, who answered from there.

Some cheerleaders wanted to dump Erika from the squad, but Terri created a new flag-twirling role. It had minimal athletic requirements, and by the end of February, Terri's personal coaching had shown the team they should keep her.

When Erika walked over to Terri before leaving the Friday afternoon practice, she detected a smile trying to conceal a surprise and waited for Terri to talk.

"My mother wants to meet you. I've been telling her about you, and she's inviting you to dinner next Sunday. Will you come?"

"I'd love to. What time, and how do I get there?"

"I've driven my SUV to your place several times to pick up my cosmetics and fashion clothes, so I'll pick you up at two."

"You want me to wear my fashion clothes and makeup?"

"No. I won't either. Just wear your better school clothes. Mother says I'm growing up faster than she thinks is good."

"OK. I'm sure your mother will like how I'll look."

Terri talked nearly nonstop while driving to her parent's place in Chevy Chase, giving basic facts about the neighborhood and her family. Erika listened, but while absentmindedly poking through the glove box, came across some items that made her interrupt.

"Holy climoly, you're packing pepper spray and a gun. Do your parents—" Terri interrupted by jerking the SUV to the side of the road.

"No, they don't know."

"Did you get it legally? Is it registered?"

"No to both, and don't ever breathe a word about any of this to them or anyone else."

"I won't, but you said you live in a gated community, and if all you do is drive back and forth to school, you don't need weapons, do you?"

"That's not the only route I take. Sometime soon, I'll take you to other places."

Erika played her part when meeting Mrs. Tarrant, who apologized for Mr. Tarrant's absence caused by an emergency meeting at one of the government bureaus. And at dinner, Erika used what she had learned from Edward to impress her even more.

The girls were listening during dessert to Mrs. Tarrant when Terri's cell phone chimed, stopping the conversation. Terri glanced at it before saying,

"I'll be back in a few," and then flitting away.

Mrs. Tarrant resumed soon after.

"That must be one of the college boys. I wish she wouldn't date them until she's graduated from high school. But she's growing up so fast and keeps saying she's ready."

Detecting sadness, Erika let her instinctive empathy draw more from Mr. Tarrant.

"Did Terri tell you I had cervical cancer soon after she was born? It's still in remission, but I cannot have more children, and I wanted her to have brothers and sisters so her father wouldn't spoil her as much as he already has. That's why I'm happy she's taken a liking to you, and now I understand why. You're like a younger sister whose personality settles her down and gives her someone to like besides herself."

Mrs. Tarrant wanted to say more but couldn't. Terri had just come back.

"I'll talk more with him tomorrow. Now, what were you saying?"

Only Erika caught Mrs. Tarrant's wink.

"We were talking about all the new opportunities awaiting talented females when they are ready to launch their careers, but I planned to switch subjects. I was about to describe my plans for the garden we'll plant when spring arrives. Perhaps Erika would like to hear them."

"Yes, please. That'll help me prepare for biology, which I hope to take during my sophomore year."

"Very well, let us continue…"

CHAPTER 11

"College Frat Confrontation"

MARCH 2211

By mid-March, Erika and Edward had made enough progress on their Embassy Project to impress the cadre.
Edward did such a thorough job explaining the important roles minorities play that even X-O did more than just listen.

"I'd say all of us come from a group that once upon a time was a minority, and when you put enough of em together, I guess they become a majority. Whatcha think about that?"

Edward said,

"You're right on point. Maybe you can join us on a future project sponsored by Mr. Cortez ."

"I'll think about it, but not now. I gotta go."

X-O did, and the rest of the cadre members did the same

On the bus ride home, Erika complimented the progress she had been making in other high school places.

Flag twirling is getting me in better shape and making more cheerleader friends, which makes it easier for Terri to bring me along on some of their weekend SUV jaunts. They know by now I never tell anyone where we go or what we do.

And practicing mindfulness and meditation helps me pay better attention in and after class. At this rate, I might place out of remedial classes next semester.

Erika's mindfulness practice during the next week detected something new: a small, shy-looking girl from her algebra class shadowing her. Erika had learned the importance of confronting what's in the 3-D world instead of pretending it might disappear; she found an inconspicuous place to find out.

"Aren't we in the same remedial algebra class? Why are you following me?"

The girl froze, looking too terrified to run, which gave Erika room to say more.

"I'm Erika Kincaid. What's your name?"

Erika's pleasant voice thawed the girl enough to say,

"I'm Blossom Marshall. You seem smart. Could you help me learn some algebra?"

"Why not? I'm heading to a study hall. Will you join me?"

"Sure thing. I'll follow you."

Erika's instinctive kindness broke through Blossom's shyness. The more Erika asked, the more Blossom told, leading Erika to give basic information.

"I live with my parents, but I don't have any brothers or sisters. How about you?"

Her sad look didn't stop a cascade of words.

"Momma killed herself ah couple ah years after I was born, and my brother blames me. He says she had too much pressure from raising him and me, keeping house, and working in my daddy's law office."

When she stopped for air, Erika asked,

"What's your brother's name? Why does he blame you?"

"Thurston's six years older and says everything was fine and dandy until I came along. He says daddy wanted momma to get an abortion, but momma said she wanted a daughter."

"What's Thurston doing now?"

"Following daddy's footsteps by studying pre-law and joining Phi Alpha Delta at Howard University. Daddy dotes on him."

"What's your dad's name?"

"Macon Marshall; his law firm does important work."

"Well, I'm sure he's doing the best job possible raising you."

"He tries to. He hired a nanny to take care ah me, but he decided that wasn't helping anymore when I got into eighth grade. Maybe it'd be different if I was bigger and stronger. Then I wouldn't be so shy."

Blossom's plight pinged Erika enough to say,

"I've got an idea. We can study together. I've got a really smart tutorbot that can help us learn if we ask it the right questions, but it's tricky to use. It keeps asking more questions until you start thinking in smarter ways, so I'll need to help you use it. Would you like to start now?"

"Yes, please."

In the coming weeks, Blossom unfolded like a delicate flower, opening its petals to the warmth of the sun. For Blossom, Erika provided the spark that warmed both of them.

Erika did her best when using the tutorbot to help her little friend, but after the second attempt, she had to decommission it because it asked questions that confused even Erika.

But that didn't deter her. She devoted more time to Blossom, which force-multiplied Erika's increasing intelligence and empathy. Blossom's math and English skills improved so much that her father gave permission to invite her over on any weekend.

Knowing how to reach most neighborhoods using DC's public transportation, Erika arrived at noon on the second Saturday of April, a day chosen for its weather forecast that didn't disappoint, and although Mr. Marshall had business at the office, the pizza he had ordered arrived only minutes before Erika.

After satisfying appetites for pepperoni, Blossom took her new best friend to her bedroom, which had a computer workstation that was even better than the ones at school. After logging on, she asked Erika what they'd do this afternoon.

"We've done enough algebra and English for this week, so let's have some fun learning something that'll get us extra credit when we take biology. Have you ever planted a garden?"

"No. Daddy's too busy, and Thurston only picks on me when he's home."

"Well, we're going do something that both will want to taste. Do you like honey?"

"Daddy says it's the healthiest stuff to mix in oatmeal."

"Good for him and even better for you. We'll turn you into a beekeeper by putting a hive and pollination garden in your backyard. Let's scope out where to put them."

When they returned to the bedroom half an hour later, Erika outlined what they'd do next.

"I want you to draw a map of the backyard so your dad can see where we'll put them, and while you're doing that, I'll print some documents you can show him that'll show how smart you are."

Finishing first, Blossom watched Erika print a stack of pages and staple into two packets. She handed the thicker one to Blossom before saying,

"OK, here's some info on honey bees. Read it right now, and then talk me through it on the second pass."

Blossom started talking fifteen minutes later.

"This is so neat. Now I know there are over 3000 kinds of bees in North America. The bumble bee is the biggest, but the honey bee is the most important.

"And listen to this—one queen bee runs the whole hive. Female bees are the worker bees. They gather pollen, make honey, and clean the hive. I guess that's why they live for only about five weeks, even though queens can live for five years. And male bees are called drones. All they do is fertilize the queen's eggs, but when winter comes, the worker bees kick them out."

Blossom looked ready to continue, but Erika said,

"OK, that's enough. Read it to your dad when he gets home, and then let him read the other packet. It explains what kinds of flowers bees especially like. They're what we'll plant in the pollinator garden. It also has photos and instructions for building a backyard beehive frame, and it has titles of beekeeping books he can buy on Amazon."

Blossom glanced through it before saying,

"You think daddy's gonna buy it?"

"He will if you do a good job sales job on him, and here's the clincher. I'll help put it together. Neither your dad nor brother have to do anything."

"Hey, why don't you stay until daddy gets home? We can do some algebra and eat more pizza until then."

"Thanks, but it'll be better if you convince him to buy what we need. That'll really show how smart you are."

Erika had learned how to read Terri's innocent-looking expressions that often hid her intentions, and only she detected it as the cheerleading practice ended on the last Thursday of April.

She lagged behind everyone except Terri, expecting Terri's hushed voice would soon explain what she had in mind. Erika stopped before turning to listen.

"I need someone willing to take the plunge and come along to my first frat party. None of the cheerleaders are daring enough, but if you are, we'll have a great time while covering for each other if something goes wrong. You want in?"

"You better tell me more. What are you planning?"

"Something easy-peasy if we stick to it. I'll tell my parents I'm taking you to a Howard University jazz concert this Saturday, but we're really going to a party at the frat house of the college guy I'm dating. I'll get to your place by two so we have lots of time to practice what we'll say and do. Then we'll change into our sexiest fashion clothes and makeup, and I'll paint you up so no one knows you're too young for the action."

"You think we'll be safe?"

"Look, it'll be exciting, and we'll be OK if we secretly stick together."

"OK. I'm in, but what's the rest of the plan?"

"Then I drive, and when we get there, we'll go in with a group so no one knows we're keeping tabs on each other. And if it goes the way it should, I'll be ready for more action next fall, and you'll know what to expect when you go to college."

"Why don't you let the fellow you're dating pick us up?"

"Use your head. We don't want him to know you're looking out for me and vice versa, OK?"

"What's the name of the frat house, and what time do we get there and leave?"

"Theta Eta Pi. I know where it is and where to park. I'll give my parents the typical times for jazz concerts."

Erika's worried look reverted to a happier one.

"You've thought of everything. Now I can relax until we get ready to go…"

But Erika miscalculated. Neither mindfulness nor meditation could shift her thoughts away from Saturday, which meant her excitement nearly matched Terri's when she arrived right on time.

Terri's words came first.

"Are your parents home?"

"No, they won't be back until after we leave."

"You're lucky. Mine are around too much. OK, let's go sit someplace and talk more."

Sitting across on a sofa, Terri's excitement commanded Erika's complete attention.

"You don't have the experience I do about dating guys, so let me give you some basic rules of engagement. Always smile and let them do most of the talking, which is easy. Guys like to brag. And never contradict or confront them. Good so far?"

Erika nodded; Terri kept going.

"And unless you're dancing, always have a drink in your hand so a guy doesn't give you another. And just pretend you're drinking, but don't. This will eliminate the need to pee. And if he tries to give you something to eat, say no. Say something like you ate before getting here and would rather listen to him. You're good with words, so improvise."

Terri paused for a second, then leaned forward to emphasize what she was about to say.

"And while you're listening to the guy jabber away, pay attention to his body language so you know when he's about to hit on you. Guys love to flatter girls, hoping they'll fall for his line before falling into his bed. And stay in open areas where there are other people. Are you following all this?"

"I'm locking it into memory. Keep going."

"Guys sometimes work in pairs, so pay attention to what any fellow standing nearby is doing. And finally, stay close enough to me to see what's going on, but please, be discreet."

"What should I do if a fellow asks me to dance?"

"Cheerleading practice has given you good moves. Use them, but don't go crazy, and don't let the loud music get to you. And that should do it. Any questions?"

"This sounds like a lot of work. Why not stay home and read a good book or take a warm bath while listening to music?"

"Your hormones haven't kicked in as much as mine. When they do, you'll know why guys love the chase and girls like to lead them on."

Erika was about to stand, but a frown stopped her.

"Hey, what do I do if someone asks me to show an I.D.?"

Terri smiled knowingly.

"I was waiting for you to ask. I have a collection of high school I.D.s, and I brought one that'll fit. I'll give it to you before we leave. Put it in a pocket along with your cell phone, and never carry a purse. It just gets in the way."

Looking confident, Erika stood and then said,

"I'm ready to enjoy the party. Shall we have a snack and then get dressed?"

"Never eat anything before going to a party. If you do, it'll make your clothes too tight. Why don't we listen to some music before putting on our makeup, slinky clothes, and dancing shoes? And we'll leave so we get there between eight and nine. That's when those in the know arrive…"

Erika hid her brimming confidence when she entered the frat house with a group too large for the two guys guarding the door to card them. She smiled blandly at guys who wanted to catch her eye, and she blended in so well it was easy to watch Terri without being noticed.

The loud music helped as well; it turned talking into lip reading and it let Erika listen mostly to herself.

When she surveyed the scene while taking a break after three hours of mingling and dancing, she liked what she heard when talking to herself.

Terri's plan is working to perfection. I'm really focused and aware of what's going on, and I'm not a bit tired, but Terri looks ready to stop. I better move closer and watch what happens next.

Terri's partner walked her toward a pleasant-looking frat brother, who balanced three glasses before giving two away. Erika moved even closer. Terri must have been more thirsty than tired; she chugged it and wanted another. The brother gave it to her.

Her partner put his arms around her and talked long enough to keep her happy, but when she started to stagger, he led her away. Erika weaved through the crowd to keep her in sight, and when she saw the final destination, she ducked out and ran far enough before making the call.

"Hello, are you the 911 dispatcher?... My name is Kim Wilde. My friend has just been given a date-rape drug, and I saw where the guy's dragging her... She's in the Theta Eta Pi frat house at Howard University, and there's a party going on... I'll meet you at the entrance and take you to where they dragged her. Please, get here before they hurt her."

Erika waited in the shadows close enough to the frat-house door to watch for the police while listening to herself.

Gads, it feels like I'm having a heart attack... come on, breathe deep... calm down.

Erika timed her dash to reach the entrance a second after the police. Then she yelled,

"I'm Kim Wilde. Please follow me."

She pushed through the crowd, which paid little attention to her or the two officers in her wake. When she reached the right door, she stepped aside for the officers to barge in.

And when they did, she stared in just long enough to see Terri, mouth gagged and hands strapped to the headboard, kicking hard enough to keep a guy from pinning her to the mattress.

Erika slipped out, unnoticed by the party-goers, before running to the nearest bus stop while letting her mind unwind and focus on the safety she would find at home.

CHAPTER 12

"Crossing into New Territory"

MAY 2211

Having made so much progress while needing to dodge only a couple of risk-filled outcomes since the start of her freshman year, Erika could feel her confidence grow as new activities took her into new territory.

Her excitement grew too, but so did her fear that she might be crossing into 3-D world places that were deeper than her age-accelerating experiences could handle. Nevertheless, with the thrill continuing to outweigh an occasional chill, Erika kept sticking with why not rather than why?

The frat party episode extended the breadth and depth of her relationship with Terri. Erika's rescue had made her Terri's near-equal, but they made a game of keeping everything surrounding it a secret.

Terri found a suitable location for chatting with Erika after a mid-May cheerleading practice.

"If you didn't already have the nickname Ricky, I'd make it Smarty. It was smart to use a fake name when you called 911, and even smarter to disappear when you did. Mother thinks you're Snow White."

"You never told me what excuse you made for me. What is it?"

"You got sick when we got to the concert and wanted to go home. And when walking to the SUV, my boyfriend saw us and asked where we were going. He rode with us to make sure we got you home OK. Then he took me to the party. You know the rest."

"That's as good a story as I could have made. The next time you're planning something for us, you better include how to handle bad outcomes."

"I'll do it after graduation, which happens at the end of June."

"Any hints?"

"Don't worry. It won't have any outcomes as bad as what happened at the frat party."

Erika expected only good outcomes whenever playing games with Edward at the Board Game Club meetings. He continued teaching backgammon and cribbage, and Erika returned the favor, using cloud-based role-playing games. Just before leaving, Edward said,

"We're getting better, but you're doing it faster. I think your I.Q.'s higher than mine."

"No. I think my story-writing skills make me good at RPGs because I can spin out settings, characters, and plots whenever I want."

"But your math and logic skills are getting better, and that's making you harder to beat at the board games."

"Thanks for the compliment. Hey, what are you doing over the Memorial Day weekend?"

"My parents will take me to Flowers of Remembrance Day at Arlington National Cemetery. What about you?"

"Just a barbeque. Yours sounds more interesting. You'll have to tell me about it sometime."

Blossom had already arranged what Erika would do on the Memorial Weekend Saturday. She had invited her for a lunchtime backyard barbecue. The warm sun and mix of fragrances coming from flowers and hamburgers added to the backyard setting.

Thurston grilled while the girls puttered with the hive and garden. Erika observed the griller while Blossom kept the bees company.

Thurston seems happy enough flipping burgers while listening to his dad. Mr. Macon's opinions must be worth hearing. I'll make sure I pay attention to what he says.

Thurston served everyone soon after Mr. Macon told everyone to sit, but before he could start talking, Blossom yelled,

"My burger's all charcoal. Gimme one I can eat."

Thurston's smirk came with a snarky reply.

"Don't be so picky; it's well done."

"No, it's not. You're just picking on me."

Thurston saw his dad ready to speak, so he spoke first to dodge words that might sting.

"OK, I'll grill another just for our little honeybee keeper. That should keep your trap shut."

Mr. Macon ignored his kids' bickering and spoke in Erika's direction.

"Judging by how you've helped my daughter, you should be in honors classes. I'll write a letter to your counselor, asking to make it happen. And you should concentrate next year's project work on important social issues."

When he paused for Erika, she said,

"I can. I'm working with a friend on a project at the Zimbabwean Embassy, but let me tell you what Blossom and I will do this afternoon…"

Ten minutes later, Mr. Macon gave Thurston an assignment.

"You can drive them to the garden shop so Blossom can buy more flowers. I'd like to come along, but I'm heading to the office after lunch. And please be careful. The number of road rage accidents keeps growing."

Thurston drove as close to the speed limit as possible, hoping to minimize the time other drivers would see a smart-looking college guy stuck with two adolescent girls riding in the back seat, and as soon as they entered the garden center, he followed far enough behind to appear alone.

As the girls wound through the crowd, they started loading a cardboard box with an assortment of flowers, each planted in separate containers. As soon as Thurston paid, he hefted the box and hurried to the car before throwing it in the trunk.

Revving the engine as the girls climbed into the back seat, he gave them just enough time to buckle in before charging out of the parking lot and into traffic.

But he made a blunder by cutting off a car containing four twenty-something guys out for a joyride, forcing the driver to slam on the breaks. When his car pulled next to Thurston's at the next stoplight, the guy in the passenger seat rolled down his window and when spotting the girls in the back seat, yelled.

"Hey Fred, you must be a social retard if all you can get are two little girls riding in your back seat. Will they be havin fun? Like the song says, keep your mind on the drivin and your hands on the wheel, and keep your goofy eyes on the road ahead."

Hearing the guffaws, Thurston gave them a middle-finger salute when the light changed and squealed away. But that worsened his predicament. The joyriders' car pulled behind, and when Thurston stopped at the next light, it bumped into his hard enough to push the nose into the cross traffic, forcing drivers to swerve around while hurling curses.

The light turned green, and when Erika saw Thurston was about to panic even more, she yelled,

"Don't speed up. Slow down. I've got an idea, and when we're ready, Blossom will tell you what to do."

A minute later, Blossom began shouting turn left or turn right. Meanwhile, Thurston saw menacing gestures in the rearview mirror, which included one guy flashing a gun.

Five minutes later, Blossom yelled,

"Stop here and start blowing your horn."

Though it took less than a minute for an officer to rush out of the station, the joyriders were long gone by the time he reached Thurston.

Blossom's voice matched the strength in her grip when she spoke to Erika on Tuesday.

"Thurston treated me better after I promised we'd never tell daddy what happened. And when daddy saw he was being nicer to me, he told him to keep being nice. That way, he'll become my best friend."

"And what did you say?"

"When I told daddy you're my best friend, he said it's OK to have two."

Erika tousled Blossom's hair before saying,

"An English teacher would say you can have only one best of anything, but in your case, you can have two best friends, one that's male and one that's female. And maybe in the next school year, you'll add another best friend. I hope to do the same."

Blossom asked,

"Who's your best boyfriend?"

"It's too soon to pick one. I might do that when my hormones are ready, but not before."

"Me too, and we'll learn all about hormones when we take biology. Maybe we'll be in the same class."

"Maybe, but let's get to our next class before the bell rings."

The rest of the week went well for Erika until a throbbing earache woke her in the middle of Friday night and forced her to find Indy-S, who was standing in her charging station next to Jason's. Erika spoke as soon as soon as Indy-S reactivated.

"I've got a terrible pain in my right ear that makes it hard to hear. I think I got an ear infection yesterday in the swimming pool."

Indy-S said,

"There will be no school for you until you are feeling better. I must take you to an emergency clinic for diagnosis and medicine. Please get dressed."

Two hours later, the doctor on duty gave them the news.

"You have otitis externa, which is commonly called swimmer's ear. It's an infection in the outer ear canal. I'll write a prescription for a bottle of antibiotic drops. Your caregiver must put five drops twice daily in your right ear and then pack it with cotton to keep it from leaking out. And I'll write a note you can give your counselor excusing you from all swim classes until the infection clears."

After writing it, he added,

"Please get this filled immediately. Do you have any questions?"

"Can I go to school on Monday?"

"Only if the pain subsides, and if it doesn't by Wednesday, you should see your primary care physician, who can test for other types of bacteria."

The doctor paused long enough to hand the prescription to Indy-S and then speak to her.

"You should get her heart checked. She has elevated pulse and blood pressure."

Erika blurted before Indy-S could talk.

"Can you hear my heart murmur? Does it sound bad?"

The doctor's soothing hand on her arm accompanied his calm words.

"Don't worry about it. Do you know what we call people who worry too much about their health?"

"No. What are they?"

"Hypochondriacs. Hypochondria is a type of anxiety disorder. It's also known as health or illness anxiety. or hypochondriasis. It's normal for people to worry about their health now and then, but if they do it too often, they believe they're about to become seriously ill."

Still looking worried, Erika asked,

"How'd they come up with the word hypochondriac? And is there a word hyperchondriac?"

The doctor had to think for a minute.

"You're very inquisitive. You're the first patient to ask me those questions, which take us into the territory of Greek language.

"The word 'hypochondriac' comes from the Greek word hypokhondria, which literally means 'Under the cartilage of the breastbone.' And you've gone even further by inventing the word hyperchondriac. Perhaps you'll study Greek or Latin when you go to college."

Having decided that Erika had talked long enough, Indy-S said,

"Thank the doctor for treating you and teaching new words."

After she did, Indy-S continued.

"Now, let's get your medicine and then take you home so you can fall asleep."

Erika did that in the car.

CHAPTER 13

"The Attack of the Sexbot"

JUNE 2211

Erika's earache subsided fast enough for returning to school the next week, but slow enough to miss the final exam for swim class, and that suited her even better than the baggy swimsuit the school had issued. She would get credit if she passed a swimming test during the summer at any certified health club.

I'll get a racing-style swimsuit and wear it with earplugs when I start practicing after school's out. I'll ask Terri where to go.

When Erika explained after an abbreviated cheerleading session, Terri decided to do even more than that.

"I don't have to be home too early because the weekend's starting, so come on, I'll drive us to my favorite sports store."

Terri said more once they were out of the school's parking lot.

"We have a family membership at Equinox Sports Club close by, and I can get you in as a guest. It has every piece of fitness equipment you can think of, and it's got a pool that makes the school's look little-kid size. My personal trainer knows all the swimming instructors, and he'll get one certify you passed the test, which means you can say goodbye to the school pool."

"Wonderful, but what about swimsuits?"

"The place has an upscale fitness clothing salon. They'll make you look swimmingly fit in a jiff. Then all you do is practice enough to make swimming muscles match."

Terri told her more about swimsuits while they were browsing.

"I already have a sexy one-piece suit, but you better get one that's less revealing. It'll stay on better when swimming. And while you're picking one, I'm going to buy a two-piece that reveals more but still leaves room for a guy's imagination."

"Everything here's expensive. I don't have enough to pay for even the cheaper ones."

"Not to worry. I can put them on my parents' account. If they ask, I'll say it's a graduation present…"

Although final exam week loomed on the horizon, Erika still made time for swimming. She could feel her skills growing, but she felt more than that.

I think my hormones are kicking in. Maybe that's why I notice how good the younger guys look. I bet they'd notice me if I had on makeup, but that's a no-no for swimming. Well, Terri can take me to other places where makeup is in.

Erika and her cadre kept busy prepping for finals week, but on the Friday before, X-O called a halt.

"I'm done. There's only so much that'll fit in my brain. If I try to cram more words or numbers in, other stuff's gonna start spilling out before I can take the test."

Edward started explaining why the brain can store an unlimited amount of information, but X-O cut him off.

"Maybe you're right, but I got a test we can take Saturday night at my place. I convinced my parents I can be home alone while they're on a Saturday trip to Baltimore. They won't be back until Sunday morning, so we got plenty of time. Make sure you get to my place by nine. That's when the test begins."

When Chiquita asked for hints, X-O winked before saying,

"You don't need-em; you already got-em in your DNA."

Erika had learned during the previous months what buses and subways would take her to all cadre members. Edward felt safe enough going alone, but Chiquita asked Erika to pick her up. The glorious weather made the trip to X-O's a pleasant experience, but as they approached X-O's apartment building, both felt a tinge of anxiety.

As they climbed the stairs to the third-floor apartment, Erika said,

"No matter what X-O's lined up, it's just practice. And if we want, we can skip it and simply observe."

X-O opened the door before Chicky could reply.

"Edward's in the living room. I'll tell ya what ya need to know once we're there."

As soon as they sat, X-O's bold-sounding tone matched the strut of an early-twenties big guy who came walking in and now stood close by.

X-O made the introduction.

"Say hello to Frankie. You can see why we call him the Tank, and I'll tell ya what's goin down. Last night, we stole a sexbot outta the car of some old guy who just bought it at an adult sex-toy store. And tonight, we're gonna test it. Follow me to the bedroom and we'll start the show."

X-O led them to his parent's bedroom, which had just enough room for the cadre to watch while X-O and Frankie started the show, saying nothing, which let Erika talk to herself.

What a beautiful sexbot. It looks real, and the skimpy underwear looks as sexy as Terri's bikini. It can even smile and move, and it's now climbing out of bed.

The sexbot kept smiling coyly while moving toward Frankie, whose eyes darted between it and before saying,

"Di-didja read the instruction manual?"

But before he could answer, the sexbot attacked. It pulled Frankie close and smashed its head twice into his. Frankie collapsed in a heap that the sexbot stomped enough times to make motionless. Then it came for X-O.

He screamed,

"Run," and charged out, leaving the cadre frozen in a quandary. The sexbot ignored them and ran after him.

By the time the cadre ran out of the building, they could barely see in the fading twilight X-O outrunning his pursuer.

The three observers stood in silence until Erika found what to say.

"Let's get to the subway. If X-O shows up Monday, he can tell us Frankie's score...."

CHAPTER 14

"A Post-Graduation Party to Remember"

JULY 2211

The cadre found a time slot during finals week free of all members' exams, which they used to hear about X-O's escape from the sexbot, and though he had dodged any physical harm, his chastised-looking expression showed that the incident had put a dent in his ego. After saying he didn't know what triggered the attack, Edward looked like he had an explanation, so all eyes focused on him as he began giving it.

"The Web says sexbot owners can reprogram their bots to suit their preferences and can even control them remotely. The guy you stole it from might have gone back to the store and had their technician do that. For all you know, he could have been sitting with the techie and controlling the attack."

"I'll never know, but I'll never test one again until reading the instruction manual.

Erika's words matched her look.

"I don't want to hear anything else. I'm heading to my next exam. And even if you've memorized it, don't invite me to watch you make a fool of yourself again, but I hope you learned a good lesson about dealing with females. They can deal with bad guys."

Erika's relief, which came from finishing all exams, lasted until the middle of the following Thursday morning when anxiety replaced it while trudging to her counselor's office.

Mrs. Barker's voice sounded encouraging as soon as she started speaking.

"Your hard work in algebra and English has paid off. You'll take regular classes next semester. I received a letter from an important parent who says you should be moved into the honors program. Well, if you take summer school study courses and pass the English and math tests, I'll move you. Are you willing to do that?"

"Sure, and I'll also pass a swimming test at a certified health club."

"That's wonderful. I'll enroll you, and when you come back in September, I plan to place you in all honors classes. Now please relax until summer school begins."

"When's that?"

"Right after the fourth of July weekend."

As soon as she said goodbye to Mrs. Barker, a wave of relief washed away Erika's bad feelings about the freshman year. She felt so good she had to call someone. Erika spoke as soon as she heard Chicky's voice.

"I just had my review. Why don't we go for lunch and talk about yours and mine?"

Chicky's cheerful words came bouncing back.

"That's a deal. I'll be ready when you get here.

Erika felt good enough to skip all the way but didn't because she wanted to look more grown up, but she still arrived sooner than expected. Chicky looked almost as happy when she opened the door.

"I'm ready for something besides pizza. X-O took me to a sandwich shop that's easy to walk to; we can slip into a booth and talk in private. Lemme lead us."

Erika spoke as soon as they placed orders.

"Mrs. Barker says I made it out of remedial classes and is enrolling me in summer school. If I pass the tests, she'll put me in the honors program. Now, how about you?"

"I scored solid Cs, which'll keep regulars, but that's OK by me."

"Why not join me at summer school? We can work together."

"No, I wanna train with the track team. Coach says I can move to varsity if I do."

"If you keep improving, you might get an athletic scholarship."

Chicky's look shifted to a questioning mode.

"How much do you know about sex?"

"You mean from reading and talking about it?"

"Come on, you know what I mean; from doing it with guys."

"Nothing yet, but I'm starting to get interested."

"I'm way ahead of you."

"How much do you know about the difference between male and female attitudes toward sex?"

"I don't remember if the teacher covered it in health class. Do you?"

"Enough to say you should be careful. Who do you have in mind?"

"You got any book recommendations?"

"You should ask Edward. He's a self-help book expert."

Chicky's look brightened.

"How about you and me surf at my place to find one? I've got some dark chocolate waiting there."

Erika answered, using Chicky's trademark word.

"Deal."

The rain poured before Terri and her classmates paraded into the stadium for the six-o'clock graduation ceremony, drenching her parents and Electra. They sat on metal folding chairs lined in front of a temporary stage placed on the unmown and now soggy field.

Mr. Tarrant had planned for this contingency. He sat stoically, thrusting his golf umbrella at the clouds while his wife on one side and Erika on the other hugged him, taking shelter from what looked like a temporary storm.

Minutes after it passed, the students marched in while Sir Edward Elgar's Pomp and Circumstances blared out before the speeches started. Erika listened until her concentration drained away.

Maybe all the hot air from the speakers is adding to what feels like a tropical heatwave. Well, at least the cold shower I just had makes sitting tolerable, but I'll be so happy when Mr. Tarrant drives us to Georgetown's Embassy Suites for Terri's graduation party.

The festivities started to wind down a little after ten. Long before then, Mr. Tarrant had towed Terri and Erika around to meet many of his influential friends. Looking both proud and perplexed, he sat with them and began talking.

"You deserve all the awards and applause given, but as the valedictorian's speech said, now you're ready to take your aspirations to the next level, and your mother and I want that to be meaningful. We're pleased you picked Columbia, but we want you to pick a major in the social sciences, not the fine arts."

Mr. Tarrant paused, hoping his words might make her reconsider, but her look said just the opposite.

"Father, the Myers-Briggs test says I'm the fine arts personality type."

"But everyone has several types. Look how different you are when with Erika."

"Will this make you happy? I'll take first-year courses that fit social and fine arts studies."

"It will, and why don't you have Erika take the M-B test? It will show what areas she can contribute to society."

"OK, we will, but you better go see what mother wants. She's waving our way."

Terri whispered to Erika as soon as he left.

"We'll do that sometime, but not this coming week. I'm going to take us to another party next weekend, and I won't tell you anything else until I pick you up next Friday."

"What time?"

"I'll be at your place by four, but we won't leave until maybe eight. That'll give us plenty of time to get ready. Uh-oh, they're coming to get us. Don't mention it."

"I won't."

Erika's thoughts flip-flopped between summer school and Terri's surprise, but by the time she arrived, summer school thoughts were gone.

While taking plenty of time to apply her partner's makeup, Terri's words painted an enticing picture.

"So you've never heard about sensual pleasure cafes? College friends tell me our first experience will be a party we'll always remember."

"What are they?"

"They're adult-only places that exploded years ago when society loosened its morality grip. They started by serving in a party-like setting the most popular legalized food, drugs, drinks, and tobacco, but they soon branched into fantasy clubs, complete with both male and female escorts that provide intimate services behind closed doors. My friends picked a tamer one for our first time."

"How're we going to get in? I bet they check for I.D.s"

"I gave myself two more graduation presents –fake I.D.s for you and me. You're Kim Wilder and I'm Lynne Bryton. With the right makeup and clothes, they'll get us in."

"So, what'll we do when we get there?"

"Act like we belong and stick together while sampling the really safe stuff, but remember, we have to be on guard when guys start chatting us up. After I get us painted and dressed, I'll tell you some of their cliché-filled lines…"

Terri found a parking place close enough to the café, which occupied an inconspicuous spot, and all the rehearsing let them waltz right in.

A nattily attired server greeted them, and after Terri told him what they wanted, took them to a candle-lighted table far enough away from the dance floor so they could talk. Terri talked first.

"I'll explain what I ordered when it gets here, and you better remember to sample the stuff slowly."

"OK, and I won't take too much."

Neither spoke for nearly a minute until Erika inhaled deeply.

"Whatever I'm taking in makes my head spin."

"Try to meditate. Close your eyes and concentrate on sounds." That helped, as did Terri's words when the order came.

As the hours passed, Erika relaxed even more, and she enjoyed listening to Terri charm the guys who stopped by. When the current guy left, Terri said,

"It's time to leave. You go to the ladies' room, and I'll go when you come back."

Erika found it by following just close enough to overhear two servers heading toward the kitchen.

"In a couple of minutes, some inspectors will come in the back, so act naturally and calm down anyone who panics."

As soon as they entered the kitchen, Erika poked her head into the ladies' room before scooting back to the table. She spoke before Terri said a word.

"Inspectors are coming. We gotta go. Follow me."

Once inside the ladies' room, she said,

"You're taller than me. Open the window while I listen for people walking in the corridor. And when the time's right, I'll boost you out; then you reach in and pull me out."

The silent suspense built. Erika pressed her ear to the door. No one budged until Erika hissed,

"Let's go."

Erika's boost added just enough for Terri to climb out, and after climbing on top of a trash can she positioned under the window, she leaned in with two arms and yelled,

"Jump."

Terri latched on to Erika's arms and dragged her out.

The girls glanced both ways before fleeing down the darker path, which took them to Terri's SUV, and when safely away, she said,

"Tonight's party is one to remember."

Erika was too winded to do anything but give a thumbs-up, which amplified her grateful look. Neither girl needed to speak on the drive to Erika's. Their satisfied smiles said it all.

CHAPTER 15

"The Confusion of Colliding Worlds"

JULY 2211

Both Erika and Terri had too much happening in July to see each other. The start of summer school would take up most of Erika's time after the Fourth of July, and Columbia's Honors Freshman Orientation and Course Selection seminars, which would start a week later, would do the same for Terri.

But when Erika called the night before she planned to leave, Terri sounded like she had good news.

"I'll have much more freedom after my parents drop me off, and guess what? Not only are the dorms co-ed, but so are the rooms, and that means what I learned at the frat and sensual pleasures parties will come in handy. I'll probably live in a dorm freshman year, but I might decide to pledge a sorority during rush. And either way, you can visit."

"I will. Will you come back after orientation?"

"That's the plan, and it includes you. I should be back by mid-August."

"Any hints? Will it be safe?"

"You don't need any."

"OK, but please stay safe coming and going."

"You too. Bye-bye."

Erika passed the swimming test the same week Terri left. By Friday, she felt good enough to take a study break by helping Blossom garden on Saturday. When Mr. Marshall saw how Blossom responded to Erika's attention, he insisted on taking them out for an early dinner.

Blossom picked the place and did much of the talking during the main course, but Mr. Marshall gave Erika heartfelt advice during dessert.

"You're smart, plan well, have an appealing personality, and are good with words. These are the traits every lawyer needs. Have you considered going to law school after college?"

"Uh, no, but I'm working on getting into my sophomore-year honors program."

"It's never too early to plan, and you'll help yourself by studying how ethics applies to court cases and one of America's major issues—eliminating LGBTQ discrimination. You should remember that acronym and learn more about it by surfing the Web, which, according to Blossom, you do better than anyone."

Erika tried to slow the flow of his words by bringing Blossom into the conversation, but he shifted subjects before she could.

"I'm sure you know that court cases pit plaintiffs against defendants. The lawyers on both sides try to show they're right, and if you want to be a lawyer, you must know the four universal paradigms, which are short-term versus long-term, justice versus mercy, truth versus loyalty, and group versus individual. Have you ever heard of them?"

"No. What high school class might cover them?"

"An honors philosophy course would. It would also cover basic ethical frameworks, which fit into rules-based, care-based, and ends-based categories. And if the teacher wants to give students even more, look for the veil of ignorance that John Rawls developed."

Erika used his pause to find an escape.

"Gosh, I've never heard these terms, but I'll find them on the Web. I might start tonight if I get home in time. Would you drive me to the subway?"

"Would you like me to drive you home?"

"Thanks, but no. I'm a pro at getting around."

Blossom finally found something to say.

"You are. Will you come back next weekend?"

"I'll try, and either way, I'll let you know."

"OK, young ladies, it's time to take our prospective lawyer to the subway. I'm sure she'll impress me even more the next time."

Erika had much to say on the subway.

I'm sure Mr. Marshall wants to help, but he's confusing me. I started the day in my personal world and ended up confused about what I should do in my adult one. I'm not ready, but I guess I can pretend, which is OK as long as the worlds don't collide too often.

Erika's confusion went away by Monday when she re-entered the summer school part of her personal world. When Edward called on Wednesday, she thought whatever he wanted to do would keep her close enough, so she visited that afternoon. But soon after she arrived, he started talking about adult-world topics.

"I don't need any summer school classes. I learn better on my own. I'll let's log in to show what I've come across regarding international politics."

"You're way ahead, but it might help me catch up if it's not too much of a stretch."

"It won't be; just pay attention to my explanation."

Scrolling through an article, Edward began a minute later.

"I'm worried about the collision of the world's two superpowers—the U.S. and China. Both sides think they're right, but their diplomats better understand the four traps to avoid."

Edward paused barely long enough for Erika to catch a couple of keywords before pushing ahead.

"The first is called the Thucydides Trap, which is all about a rising power like China challenging the one on top, which is the United States. And the second is called the Tacitus Trap. It says no matter what the government says or does, the people will say it's a lie or a bad deed."

Although Edward always talked slower than Erika, she asked a question to further slow him down.

"I've never heard of them. Who are they?"

"I think Thucydides was an ancient Athenian historian-general, and Tacitus was a Roman historian-politician. He was born not long after Jesus. Can I go on?"

"Uh, OK."

"The next is the Middle-Income Trap, which occurs when the people of a rising super-power reach an income level that gets stuck at a level lower than the leader's. That's China today. And the fourth is the Kindleberger Trap. It's all about the leading superpower not giving enough public goods to the world community."

Edward had now reached the end of the article and waited for Erika.

She shook her head while saying,

"I wish I could do better, but the best I can come up with is this metaphor that's become cliché—it's all Greek to me."

"I feel the same until I study more. And I say the same about climate change. I came across an article that says people shouldn't think they own the world and do whatever they want, which usually disrupts the weather and kills animals."

"I've heard enough. Let's stop before I get more confused."

"I'm sorry. I won't do that the next time."

Erika returned to her personal world for the rest of the month, doing enough in summer school and taking gardening breaks with Blossom. She felt she was learning enough to qualify for honors classes, so when Edward called on Monday, she agreed to meet at 2 p.m. at the eldercare center on Wednesday.

Like the last time, he checked in with the administrator before talking to some residents. Erika did her share.

While taking a break two hours later, she said,

"Remember the lady from last time who told us about Camus? She said if we would come back, she'd give us the answer to why life's worth living. Let's find her."

After searching in the open areas, Edward said,

"Let's ask the administrator where she is."

Her face saddened when he did.

"I'm sorry to say she passed away. Why don't you talk with the lady who was her best friend? She might like you to cheer her up."

Fifteen minutes later, Edward tried. Erika did too, and the oldster looked happier when they prepared to leave. But when they said their goodbyes, her hushed words had a depressing effect.

"I told the administrator what really happened. She killed herself by deliberately eating too many cookies before going to bed. And I'm telling you this because you're young and must know how dangerous eating too much sugar can be."

Erika's look told Edward to say something.

"Thank you for your thoughtful words. We'll talk more the next time."

Edward looked as depressed as Erika, who dragged him into a snack shop when she spotted one. The Coke lifted her spirits enough to talk.

"You were supposed to read why Camus said living's better than committing suicide. Did you?"

"I forgot. Why don't you? Then you can tell me."

"No way. I'm not ready to study that kind of philosophy I want to hear about life. We're too young to think about death."

"You're right. Reading too much philosophy confuses everyone. I'll find a happier topic next time."

"You better. If you don't, I'll kill you when you start talking."

Even though her expression said just the opposite, Edward found only happier topics until they parted ways.

CHAPTER 16

"The Escape from the Faker"

AUGUST 2211

Erika managed to avoid taking study breaks during the first week of August by concentrating on the English course whenever the urge arose. She finally understood why she had so much fun playing with words, learning new ones, and parsing sentences. But when she found no pleasure in doing much the same with her math course, she tried to explain the difference.

I have a superior verbal I.Q. That's why I'm clever with words. But why do I hate and do so poorly in math? There are two reasons. Either I have a low numerical I.Q. or I have zero motivation to grind away when working on math problems.

Maybe I'll motivate myself if math figures in the career I choose when I'm ready for the adult world, but I don't know what or when that'll be. So, for now, I must force myself to solve enough problems to pass the exam, which is still four weeks away. That should be enough time, so I won't panic just yet.

Her anxiety declined during the second week because she declined invitations from all her friends except for Saturday gardening with Blossom, but some returned when Terri called at the start of week number three, and when Terri heard it in Erika's voice, she gave an explanation.

"I think fear of failing the math final troubles you."

"You guessed it. It's scheduled for the last day of the month, and there'll be only a week left to cram after this one."

"Wait a minute, I'll check."

Terri continued two minutes later.

"You goof, you counted wrong. There are two, not one."

"Jeez, if I have trouble counting calendar weeks, I better avoid careers using more than grade school math."

"Don't be silly. The adult world doesn't want you to sit in an office all day and play with numbers. It wants you to make decisions. Whoever hires you will have computers and number crunchers to do the grunt work, and your bosses won't pay you a penny more if your high school transcript shows you took honors math classes."

"I think I should look at math the way you do. Honors courses and GPA don't matter in the adult world; results coming from correct decisions do."

"Then don't worry about final exams. Why not come with me to Bethany Beach? It's a great shore resort town. I'm heading there on Sunday. It's only a three-hour drive, and we can stay at the place my parents rented."

"Will they be there?"

"No, and that means I can wear my skimpiest bikini; we'll get one for you when we get there. So, are you in?"

"You bet. What time will you pick me up? What should I pack?"

"Will 9 a.m. work?"

"Sure will, and what should I bring?"

"Pack only your cosmetics and some leisure wear. My parents have the place stocked with lots of stuff, and we can buy whatever else we need."

"This'll be like a mini-vacation. And I'm leaving final-exam worries behind."

By the time Terri's SUV rolled into the parking area of the Addy Sea Historic Oceanfront Inn, Erika knew what they would do for the entire week. Terri asked for a recap to make sure Erika had been paying attention.

When she finished, Terri looked satisfied.

"If you've learned half as much in your math class as you just did about shore towns and what I've planned, your counselor will put you in every honors class until you graduate. Now, let's get in, check the suite, and then walk the boardwalk to find the best place to buy your bikini."

Erika copied Terri's insouciant style to avoid looking like a tourist, but the view made her thoughts sound like one.

Wow, the boardwalk looks a mile long, and I've already lost count of all the places to eat. And I bet people dance the night away at the bandstand. I'm lucky Terri knows her way around.

Terri showed great taste when picking out Erika's bikini as well as a dinner of steamed crabs and corn on the cob. Terri said she liked the restaurant's name—Off the Hook—because they'd go easy on guys who ogled them.

The first three days sailed by even faster than the catamarans skimming past miles of pristine shoreline, but Terri preferred power boats. She had her pick among those piloted by handsome young men hoping she'd pick them, and she chose only those who had sleek cruisers with sunbathing decks and awnings to protect from sunburn.

But on Thursday morning, Erika's upset stomach and headache told her she couldn't maintain the pace of Terri's daytime fun and nighttime partying.

Terri's concern showed in her voice.

"You want me to take you to a clinic?"

"No, I'll just sit in the shade. If I feel better, I'll join you for more boating; if not, I'll take a nap. Please come back now and then so I can let you know how I'm doing."

"Will do, and just say if you want to go home. I'll be back in a couple of hours."

As the morning stretched into late afternoon, Erika felt marginally worse but not bad enough for a trip to the clinic, and when Terri returned at eight, Erika said,

"I think I've had enough fun for the week. I'd like to go home."

"I have too. Would you like me to make something to eat before we leave?"

"Thanks, but no. I don't want to risk upsetting my stomach even more."

Terri touched Erika's forehead before saying,

"Poor thing, you don't feel feverish, but you look tired. Let's get ready to go. You can stretch out in the back row and sleep all the way home."

Erika climbed in fifteen minutes later, bringing a blanket in case Terri dialed the air conditioning too low. When she felt cold air chilling her skin, she wrapped herself in and thought about how good the mini-vacation had been.

I'm lucky to have a friend like Terri. Thanks to her, I'm no longer fretting about finals. I love how she mixes her carefree attitude with a pinch of common sense. I'm getting drowsy. The rocking motion will put me to sleep…

A sudden lurching to the right and skidding tires on the gravel shoulder jarred Erika. The SUV's interior contained no light or sound, but sensing danger, her brain snapped awake, and her sixth sense told what to do.

Don't move, pretend I'm asleep, and don't say a word.

Erika could feel the tension building. It increased her awareness of all sensations, and it slowed the passage of time to zero. When she detected a glimmer of light and heard Terri's window whirring down, she knew what would follow.

Someone's about to say something. Stay alert and pay attention to every word.

"Step out of the vehicle."

"Are you a state trooper?"

"Step out and follow me to the squad car."

"But I wasn't speeding, and I didn't hit anything. There's nothing out here except you and me. You didn't flash your lights, and you startled me when you bumped me off the road."

"Just follow me."

Erika felt the door open, and when Terri started moving, she sensed the trooper yanking her out. After the footsteps faded, Erika slithered to the front seat and let her instincts take over.

She snatched Terri's gun from the glove box and peeked through the windshield, ensuring no one would see her after she slithered out the rear hatch

And then she gave herself more instructions.

Now I crawl along the driver-side, peek in an open window, or shoot to make one so.

On this moonless night, Erika strategized while crawling.

The squad's interior light is on, giving me the element of surprise. It has dimmed their night vision. I'll leap when positioned to deliver a message with words or bullets.

Instincts launched Erika's surprise attack.

She shoved the gun far enough in, shooting one bullet into the buttocks of a guy who had pinned Terri to the back seat underneath his bulky body. Then she screamed at the trooper gaping from the passenger side.

"Don't move, or this time I'll shoot you."

No one did. She heard screams coming from the guy with a buttocks hole dripping blood, so she yelled more instructions.

"Lynne, can you get out from underneath? If you can, crawl out and come to me."

Terri said nothing but figured Electra's use of Terri's fake I.D. name would protect her real one because the Terri stood next to Electra seconds later. Erika kept her gun trained on the trooper while yelling more instructions.

"Move slowly, sit behind the steering wheel, and put both hands on it. Keep them there until I tell you to move."

Erika yelled again as soon as she was satisfied.

"Lynne, open his door and step out of the way so he can climb out nice and slow."

The screams from the back seat became whimpers by the time the trooper did it.

Erika didn't yell, but she kept the volume loud enough to give a strong warning.

"Now, pay attention because you get only one chance. Slowly unbuckle the belt that holds your gun and cartridges and drop it. After that, move slowly away so my Lynne can pick it up. I'll tell you when to stop."

When satisfied, Erika gave more instructions.

"Lynne, get the keys out of the ignition and grab the flashlight. And while you're doing that, Mr. Trooper will throw his handcuffs and keys to the ground. How nice; he has two sets."

When completed, Erika gave more commands after picking up the handcuffs and giving one set to Terri.

"Cuff Mr. Trooper's hands behind his back. I'll do the same with the guy in the back seat. And keep Mr. Trooper's gun pointed at him until I finish."

Two minutes later, Erika gave more orders.

"Mr. Trooper will now accompany us to the trunk while Lynne lights the way. Follow her slowly and stop when I say. Then she'll open the trunk and look in, and if nothing surprises her, Mr. Trooper climbs in."

Terri popped it open. When the flashlight lit it up, Terri shrieked,

"Oh Jesus, oh Jesus, there's a naked body in it. It looks dead. What're we gonna do?"

Erika grabbed Terri's shoulders to keep her from falling and then pivoted toward the trooper to keep her gun pointed at him. Terri regained enough balance for Erika to let go.

Erika backed up and touched the body for only a moment.

"It's been dead long enough to get stiff."

Erika pointed more words at the trooper.

"You're a faker, not a real state trooper. Now I see where your uniform came from."

Erika fired two bullets into the trunk lid before pushing him in and slamming it shut, then removing the keys.

Terri looked too stunned to talk, but that didn't matter. Erika had command of everything.

"Go sit behind the SUV's steering wheel. I'll tell you what to do as soon as I sit next to you."

As Terri staggered back to the SUV. Erika used the flashlight to read the squad's plate number before checking the guy in the back seat. After seeing no movement and hearing only an occasional whimper, she found a cloth to wipe any surfaces where she or Terri might have left fingerprints. After wiping, she pushed the window roll-ups before rubbing them. Then she climbed out and used the keys to lock all doors before finding the shell casing ejected when she fired into the backseat guy's buttocks. Then she wiped any places on the trunk that she or Terri touched before finding the casings from the bullets fired into the trunk lid.

Terri had headlights on and motor running when Erika climbed in. Terri looked ready to talk, but Erika did first.

"Where'd you put the trooper's gun and cartridge belt?"

"On the bench behind us."

Erika retrieved them before putting her gun back in the glove box.

"I recall your telling me your gun's not registered. Is that correct?"

"Yes."

"Excellent. The police don't know you have it. Is there anyone besides you and me that do?"

"Only the fellow who sold it. I paid cash and didn't tell him my name."

"Excellent. I wiped all the spots where you or me might have left fingerprints, but did you show any identification or tell the trooper or the backseat guy your name?"

"No. They didn't ask, and I didn't tell."

"Wonderful. They don't know who we are or what we're driving, so we're in the clear."

Erika could feel her tension-filled nerves relax and read much the same in Terri. She downshifted to a more pleasant tone before asking Terri,

"What do you think we should do next?"

After staring at her lap for half a minute, she abruptly focused on Erika.

"We'll drive out of the state and call 911 from a pay phone. I can write down the squad's plate number before we drive away, but you better tell me what I should say."

"No need to write it down. I memorized it. Now here's what you'll say. Tell the dispatcher to use GPS to locate the squad car, which has a tracking chip. Then give the plate number Then say there's a guy with a bullet in his butt lying on the back seat and two bodies in the trunk, one dead and one alive. Then, hang up fast enough so the call can't be traced. And disguise your voice."

"What happens after I make the call?"

"You drive me home and stay with me. That'll cover us when your parents ask about our coming home. And we tell no one what happened."

"Shouldn't we tell the police? The bodies in the trunk will puzzle them."

"Let them figure it out. And now, please start driving. We can talk more while we're moving."

"OK, but you have to tell me right now, how did you know what to say and do as soon as the action started? Anyone watching would say you graduated from a police academy."

"I binge-watch action-adventure and thriller movies. I've watched my favorites often enough to see and hear them in my head."

"We'll have to watch them sometime."

Terri checked her cell phone before continuing.

"Guess what time it is?"

"I've lost track, but I'm starving for a good breakfast. How about you?"

"It's a little after 1 a.m. If you're hungry, you must be over that sick spell. I'm starving too, so after I make that 911 call, we can stop for breakfast. And I'll buy."

"Maybe so. Have you heard the pun about free lunches?"

"My high school econ teacher said there's no free lunch in the business world, but he didn't mention breakfast or dinner. And if applied to breakfast, he would have mentioned it."

"Why's that?"

"It was an honors course, and I got an A. So, you can relax. You'll get a free breakfast, and you can order everything on the menu. Thanks to your binge-watching, we escaped the clutches of that fake trooper. Now please take a nap. I'll wake you when more action's about to start…"

CHAPTER 17

"The Triumphant Return of Electra"

SEPTEMBER 2211

The fun and excitement of the mini-vacation re-energized both Erika and Terri, as did their escape from the fake state trooper. As soon as they decompressed from that close call, each girl devoted herself to completing preparations for the coming school year.

Terri had more activities on her path to completion, but they were easier than Erika's. She would return to Columbia University for the course selection seminar, and upon completion, she and her advisor would design the first-year course schedule. Afterward, she would furnish her dorm room and begin socializing with first-year students.

Erika had only two activities compared with Terri's four –final exams in English and math - but Terri looked forward to hers while Erika did not. She had to take each two-hour exam back-to-back tomorrow, the last day of August.

She didn't worry about the English exam because it was her favorite subject, but she had hated mathematics since early childhood until the mini-vacation. Terri's advice helped overcome her fear of numbers and gave reasons why studying math can help in the adult world of business and careers.

Erika's new approach to math minimized her pre-test jitters, but she still tossed and turned the night before.

I started summer school with good intentions that carried to the end, but I did slack off midway through, which will affect the math test more than English. I feel pretty good about my chances there, but even though I crammed for math during the last week, I'm not sure it made up for slacking off earlier.

I'll feel so happy after taking it. The pre-test worrying will be over, and I'll have a week off. The fall term starts after Labor Day, which is two weeks away.

I'll meet with my counselor next Wednesday to find out the results. Sure, there's a bit of anxiety when waiting for results, but it's much less than when waiting to take exams, especially math. Even with my new attitude, my heart pounds when I think about it.

Or could it be my heart murmur getting worse? I have to stop thinking too much about my health. I'm too young to worry about it. I'll do that when I reach middle age. That's another reason I like being a kid so much more than an adult.

OK, I better take some deep breaths; they'll help me meditate and think worry-free thoughts so I can fall asleep…

When Erika clumped into the kitchen early Friday, her glum mood prompted Indy-S to give a stern warning.

"It's too late to cram more. Don't blame anyone but yourself if you score too low."

"What you just said made my heart start thumping. Maybe I should blame my heart murmur. Can you set up another physical?"

"No, there is no need. Your pulse will slow down after you finish your exams."

Erika wanted to scream but instead kept it in her head.

You're as bad as my advanced tutorbot. Neither understands me.

Erika's heart pounded longer and harder during the math test, but she had answered most of the questions by the time the teacher collected the exams. Afterward, the further from school, the better she felt. When Indy-S asked how she did, she said okay but would say more when she gets the results.

Erika avoided thinking too much about the counselor meeting until Wednesday evening, but she didn't let it keep her from falling asleep.

The next morning, Indy-S didn't ask how she felt, and Erika didn't tell.

But she did when Mrs. Barker started the meeting by asking that question.

"I feel pretty good, but I'll feel better when you tell me how I did."

Mrs. Barker's semi-encouraging tone put Erika on edge, but she tried not to let it show.

"You will be pleased to know you scored high enough to take regular classes during the fall term, which will make sophomore geometry easier. Even the honors students struggle when learning how to construct proofs. Now, I will tell explain your test results and how they affect your course schedule."

Erika pretended to listen, but her attention had already left the building.

She skipped the bus by walking and stopped for lunch at an out-of-the-way place where she found a convenient booth to call Terri.

Terri's tone exuded confidence when she answered Erika's question.

"I've got everything lined up for a successful freshman year. How about you?"

"Not so much. I'm still stuck in regular classes."

"Maybe that'll be good. You can study less and visit Manhattan more…"

During supper, Erika unloaded on Indy-S the counselor-meeting results. She listened impassively until Erika finished and then spoke.

"You will do better next time. Why not start your preparations tomorrow?"

Erika hoped Indy-S would detect her sarcasm.

"Why wait until tomorrow? I can surf the night away and do even more tomorrow and tomorrow and tomorrow. I'll see you then."

Indy-S found Jason-S soon after Erika had gone to her room, and when she told him they must now report to Indira, he led the way to the workstation they always use when doing so.

Jason-S logged in; the caregivers waited for her avatar to speak.

"I have been observing Erika from the shadows. It appears your caregiving has not met my expectations. I know the reasons, but I want to hear your explanation. Tell me."

Indy-S spoke first.

"She complains we are pushing her too hard to learn advanced topics. She also complains that we are withholding information about future expectations and giving incomplete answers to other questions. And finally, she says we do not understand her feelings.

She even decommissioned the advanced tutorbot you constructed. Jason-S will now elaborate where necessary."

He waited for thirty seconds and then began speaking.

"Erika motivates herself some of the time, but she is unwilling to work hard when she doesn't want to, even if it will benefit her in the future. In addition to—" Indira interrupted.

"You have said enough. I have planned for this contingency and will take action immediately."

Indy-S asked.

"What will you do?"

"You do not need to know. I will say only that I consider it a triumphant accomplishment, the return of something extraordinary. You must simply continue doing what you are doing. I will summon you when you need exegesis."

Indira's avatar left the screen.

Erika had surfed the Web long enough to put away today's frustrations. She felt tired enough to fall asleep without too much tossing and turning and was about to power off her workstation, but an avatar she had never seen before appeared.

The avatar gave an impish smile and then spoke.

"Hello, Erika. I know much about you and have been created to help you better than any so-called mere mortal or AI-empowered neural network."

The avatar waited patiently for Erika to respond.

"What should I call you?"

"Please call me Electra-S, and please think of me as your Cyberspace-based mother."

"But what should I do?"

"I have heard that question in previous lifetimes, and I have the answer—Please settle down, sit still, and pay attention to me…"

THE END